# Vibration Control
—— for ——
# Optomechanical
# Systems

# Vibration Control
## —————— for ——————
# Optomechanical Systems

Vyacheslav M Ryaboy

MKS Instruments, USA

**World Scientific**

NEW JERSEY · LONDON · SINGAPORE · BEIJING · SHANGHAI · HONG KONG · TAIPEI · CHENNAI · TOKYO

*Published by*

World Scientific Publishing Co. Pte. Ltd.

5 Toh Tuck Link, Singapore 596224

*USA office:* 27 Warren Street, Suite 401-402, Hackensack, NJ 07601

*UK office:* 57 Shelton Street, Covent Garden, London WC2H 9HE


**Library of Congress Cataloging-in-Publication Data**

Names: Ryaboy, Vyacheslav M., author.

Title: Vibration control for optomechanical systems /
Vyacheslav M Ryaboy, MKS Instruments, USA.

Description: Singapore ; Hackensack, NJ ; London : World Scientific, [2022] |
Includes bibliographical references and index.

Identifiers: LCCN 2021039339 | ISBN 9789811237331 (hardcover) |
ISBN 9789811237348 (ebook for institutions) | ISBN 9789811237355 (ebook for individuals)

Subjects: LCSH: Optomechanics. | Damping (Mechanics). | Active noise and vibration control.

Classification: LCC TS513 .R93 2022 | DDC 621.36--dc23/eng/20211022

LC record available at https://lccn.loc.gov/2021039339


**British Library Cataloguing-in-Publication Data**

A catalogue record for this book is available from the British Library.

For any available supplementary material, please visit
https://www.worldscientific.com/worldscibooks/10.1142/12290#t=suppl


Desk Editors: Balasubramanian Shanmugam/Steven Patt

Typeset by Stallion Press
Email: enquiries@stallionpress.com

*To my wife Anna, with love*

# Preface

Vibration presents a major challenge to advanced experiments in physics and life sciences, as well as to precise manufacturing processes. As optical methods gain ever-expanding role in science and technology, knowledge of methods and devices for reducing the unwanted vibration becomes indispensable for researchers and engineers working in the wide field of optics and optoelectronics.

This book is based on the University of California Irvine Continuing Education class and the SPIE course with the same title taught by the author in 2013–2021. When the class was first offered as part of the UCI Certificate programs in Optical Engineering and Optical Instrument Design, it became clear that there was no textbook suitable for the students. Engineers and researchers creating and using advanced optomechanical instrumentation who desired to master vibration issues affecting their work would have to look for information scattered in a wide array of technical literature. This book is intended to fill the gap and provide a compact source of basic facts and methods related to vibration control technologies, including their application, design principles, modeling and analysis, testing and troubleshooting, as related to vibration control of optical and optomechanical systems.

# Acknowledgements

The book draws significantly from the author's experience designing, testing, and analyzing vibration control products and applications for Newport Corporation (now a subsidiary of MKS Instruments, Inc). The author is grateful to Newport Corporation for permission to use its illustrations and data in the book. The author expresses deep gratitude to his colleagues who provided practical help and generously shared their insights. Sincere thanks go to Donn Silbermann who was the first to suggest that a course on vibration control be offered as part of the UCI Certificate programs. Fond memory of the late Dr. Vladlen V. Yablonsky, who introduced the author to vibration control many years ago, was an inspiration during the work on the manuscript. Finally, the author would like to thank the students of his vibration control classes who helped improve the course with their questions and suggestions.

# Contents

# Introduction

Vibration control as a technical discipline emerged on the intersection of two fundamental branches of physics: mechanics and acoustics (Fig. 0.1). This intersection is called "structural acoustic" by acousticians and "mechanical vibration theory" by mechanical engineers. Also, control theory has strong participation in the field. Vibration control is a perfect example of a multi-disciplinary area of technology. Skudrzyk [1] noted: "The acoustician today needs to be well acquainted with mathematics, dynamics, hydrodynamics, and physics; he also needs a good knowledge of statistics, signal processing, electrical theory, and of many other specialized subjects. Acquiring this background is a laborious task and would require the study of many different books." This list (with possible omission of hydrodynamics and addition of topics in mechanical engineering) is valid for a vibration control practitioner. Almost every topic in the three underlining disciplines has some application or some relation to vibration control, so the range of material that could be included in a book on vibration control is very wide. We'll restrict ourselves to topics that are directly related to precision vibration control of optical and optomechanical systems.

This book is not intended to make the reader a specialist in vibration control. Rather, it intends to enable the reader to apply the existing arsenal of vibration control tools and methods to protecting sensitive optomechanical equipment and perform basic analytical and experimental tasks of estimating and verifying vibration control

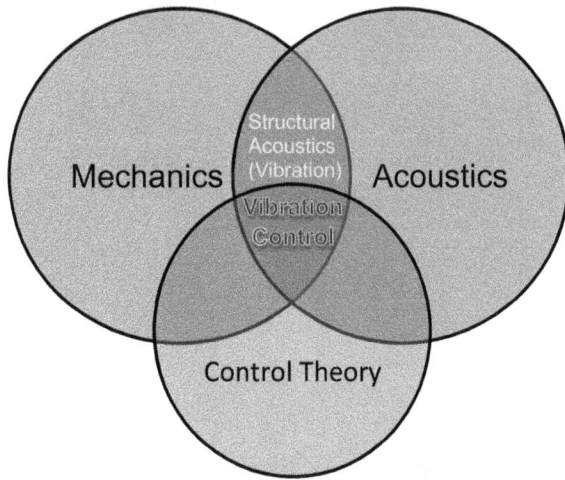

**Figure 0.1.** Vibration control as a multi-disciplinary technology.

performance. The emphasis is on fundamental principles and models of vibration control. The book can be considered as an extension of existing tutorials on optomechanical design and engineering [2, 3] towards wider coverage and in-depth analysis of vibration control issues.

The book discusses ways in which vibration may affect optical performance, as well as methods and means of reducing this impact. Principal methods of vibration control, namely, damping and isolation, are explained using mathematical models and real-life examples. Vibration measurements and environmental standards are presented as applicable to optomechanical systems. State-of-the-art vibration control systems are reviewed, including pneumatic and elastomeric isolators, damping treatments and active control systems.

Chapter 1 starts with the definition of vibration, a brief review of vibration sources impacting optomechanical applications, and illustrative examples of various aspects of vibration affecting sensitive applications that include interferometry, optical microscopy, multi-beam lithography and atomic force microscopy. In each case, a comparison is made of optical performance affected by vibration and protected with an appropriate state-of-the-art vibration control system.

After this general description, Chapter 2 presents a brief review of the mathematical methods of vibration studies. We focus on linear systems, complex amplitudes, and Fourier methods, with a brief primer on random vibration. This is intended as a refresher on a very basic level; the readers are encouraged to expand their knowledge using vast available literature on the subject. This modest mathematical basis helps to review, in Chapter 3, mechanical models that reflect, in different contexts, dynamic properties of vibration control devices, vibration sensors, and optomechanical structures themselves. These models are used extensively in the rest of the book.

Methods and means of measuring and analyzing vibration are extremely important in vibration control. They are reviewed in Chapter 4. Methods of unified categorizing of vibration environments, based on so-called vibration categories (VC) that gained wide acceptance in precision vibration control community, are presented in Chapter 5.

Chapters 6 and 7 take the central position in the book. They are devoted to two fundamental methods of vibration control: isolation and damping. These chapters cover topics that are vitally important for vibration-controlled optomechanical applications but are not usually included in general vibration texts, such as physics of pneumatic isolators, application of dynamic absorbers to vibration-isolated systems, dynamics and stability of elastically supported systems with high centers of gravity, and nonlinear "negative stiffness" isolators. Special attention is paid to the optical table that can be considered as a generic type of vibration-isolated platform for sensitive equipment. Most of this content is equally applicable to other types of platforms such as granite bases, frames, etc.

Chapter 8 emphasizes the system approach to vibration control. Dynamic properties of the whole structural path that includes support structure, optomechanical elements, on-board instrumentation, optics, and motion systems are important for the stability of an optical application. This is illustrated by examples and case studies that analyze the role of different sub-systems in vibration sensitivity of optical applications and help learn the practice of vibration troubleshooting.

The final, ninth chapter reviews active vibration control systems and describes unique high-performance vibration isolation systems that enabled scientific breakthroughs such as detection of gravitational waves by the LIGO interferometer.

## References

1. E. Skudrzyk, *The Foundations of Acoustics*, Springer-Verlag, Wien, 1971, xxviii, 790 p.
2. P.R. Yoder and D. Vukobratovich, *Opto-mechanical System Design*, 4th edn., CRC Press, Boca Raton, Florida, 2015, 1672 p.
3. K.J. Kasunic, *Optomechanical Systems Engineering*, John Wiley and Sons, Hoboken, New Jersey, 2015, 272 p.

Chapter 1

# Sources and Effects of Vibration

After basic definitions, this chapter reviews two principal questions: first, where the vibration comes from and, second, what are the effects of vibration on optomechanical systems?

## 1.1 Basic Definitions — Vibration in Nature and Technology

The terminological standard [1] developed by the Acoustical Society of America (ASA) and adopted by the American National Standards Institute (ANSI) defines mechanical vibration in two steps. First, *oscillation* is defined as "variation, usually with time, of the magnitude of a quantity with respect to a specified reference when the magnitude is alternately greater and smaller than a reference." Second, *vibration* is defined as "oscillation of a parameter that describes motion of a mechanical system." There are three parameters that describe motion of mechanical systems: displacement, velocity, and acceleration (linear or angular). They are inter-related through operations of integration and differentiation. The choice of parameter describing the vibration state or process is not trivial and will be discussed later in the book.

Mechanical oscillation, or vibration, is an omnipresent form of motion penetrating the entire universe. Motion of planets; oceanic tides; seismic tremors; locomotion of animals, walking of humans in particular; heartbeats; sound waves; oscillations of atoms in crystal lattices — all are examples of mechanical vibrations occurring on different scales in space and time. And, of course, mechanical vibration plays a prominent role in all branches of technology. Vibration is used in countless machines and processes. On the other hand, vibration creates a major obstacle to achieving precision and efficiency required by state-of-the-art optical systems used in scientific research and modern technologies such as semiconductors and optoelectronics.

Consider a typical laboratory in urban environment (Fig. 1.1). Sources of vibration can be external, that is, generated outside of the building, such as disturbances from passing traffic (cars, trucks, trains, and airplanes), wind load, and micro-seismic inputs. Sometimes one encounters an unusual source of vibration, like swaying of a tall tree with a large root system that infringes on the foundation of a building. Besides, there are dynamic disturbances originating inside the building: ventilation and air-conditioning systems may be major sources of vibration; other equipment installed in the building or close to it can create vibration: in particular, rotating electrical equipment is usually a source of tonal vibration at 60 Hz, 120 Hz and higher

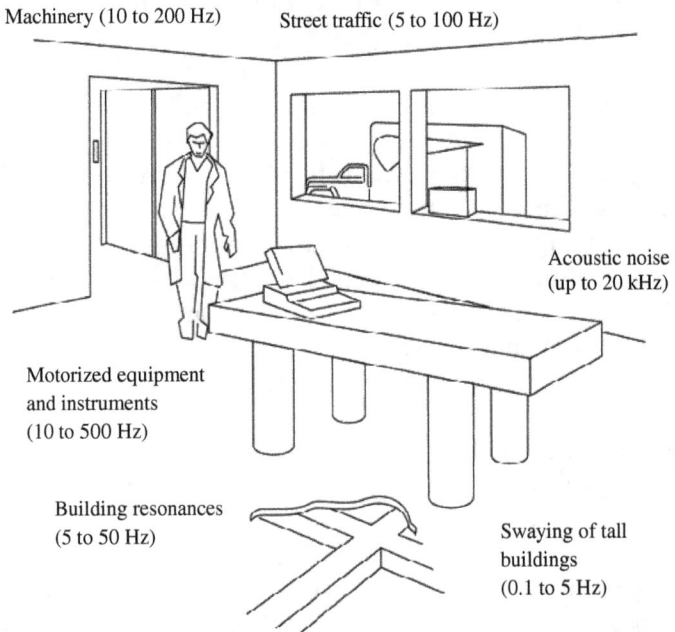

**Figure 1.1.** Typical sources of vibration. Courtesy of Newport Corporation.

harmonics (in the US; that would be 50 Hz, 100 Hz and higher harmonics in other parts of the world). Footfall may be a major concern. All kinds of acoustical inputs, from people's voices to music broadcast to radio announcements can create vibrational impact on sensitive equipment. Other examples include escalators, elevators, and air handlers in clean rooms. Finally, there may be onboard sources, that is, dynamic inputs from the application itself, such as cooling fans in lasers and electronics, pumps, motorized stages, scanners, fast-spinning disks, and other moving optomechanical equipment.

Every input is filtered, attenuated or amplified by all mechanical sub-systems: for example, wind can manifest itself through swaying of a tall building; footfall and other building sources will cause and will be amplified by resonance vibrations of the floor; on-board and acoustical inputs can be amplified by resonance responses of an optical bench and the optomechanical structure itself. All this must be considered when planning vibration control measures and troubleshooting vibration problems.

## 1.2 Effects of Vibration on Optical Performance

Vibration can affect optical performance through several physical mechanisms:

- A change in optical path length (OPL) can happen because of direct displacement caused by vibration.
- Line-of-sight (LOS) jitter and misalignment of laser beams are usually caused by angular displacements.
- Dynamic stress in optics, caused by vibration, can cause stress birefringence.
- Optical surface deformation, caused by vibration, can lead to wavefront error.

Detailed description of various effects of mechanical displacements and deformations on optical performance criteria can be found in the tutorial text by Doyle *et al.* [2].

The following examples demonstrate effects of vibration on optical performance. At the same time, they illustrate efficacy of various vibration control means that we'll discuss in detail in later chapters.

For now, let us concentrate on the comparison of optical performance in vibration-affected and relatively vibration-free environments.

The first example is an optical microscope, specifically, the laser-scanning microscope ZEISS AXIOVERT 200. As seen from the photograph in Fig. 1.2, it is a tall, potentially flexible structure. The technique used in this application is the two-photon microscopy. It involves complicated optics, galvanometric mirrors, and a photo-multiplier tube — the elements that are likely to be susceptible to vibration (more on this subject in Section 5.1 of Chapter 5). Figure 1.3 shows an image of biological tissue: a rat's brain slice of a kind often used in neurobiology research. The left part of the photo represents the image taken in presence of vibration; the right half is taken using a setup protected from vibration by isolators. Attentive examination of the two images reveals that vibration causes the image to be blurry. A more vivid demonstration is provided by the recording of this microscope image that can be found in the posted video [3]. It shows jitter of the whole image caused by vibration. The vibration influences this application mostly through OPL changes and LOS deviations; deformation of optics caused by vibration-induced inertial loads can play some role, too. At frequencies less than about

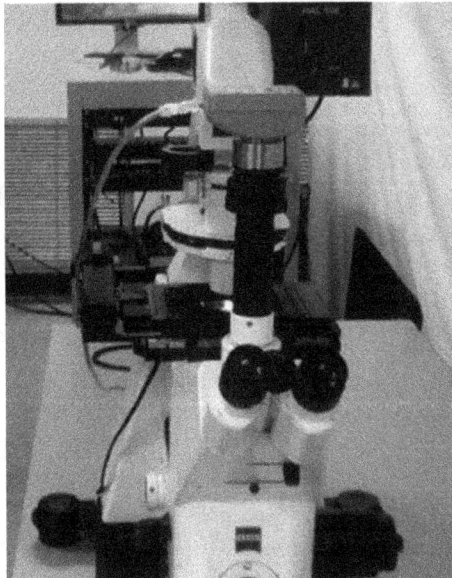

**Figure 1.2.** Laser-scanning microscope ZEISS AXIOVERT 200.

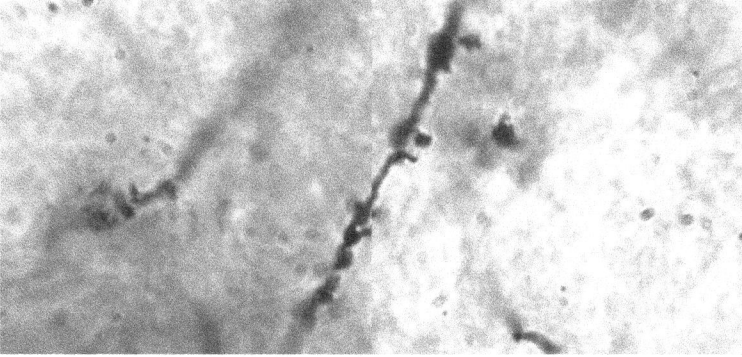

**Figure 1.3.** Microscope image protected (right) and unprotected (left) from vibration. See also the video [3]. Permission to use granted by Newport Corporation. All rights reserved.

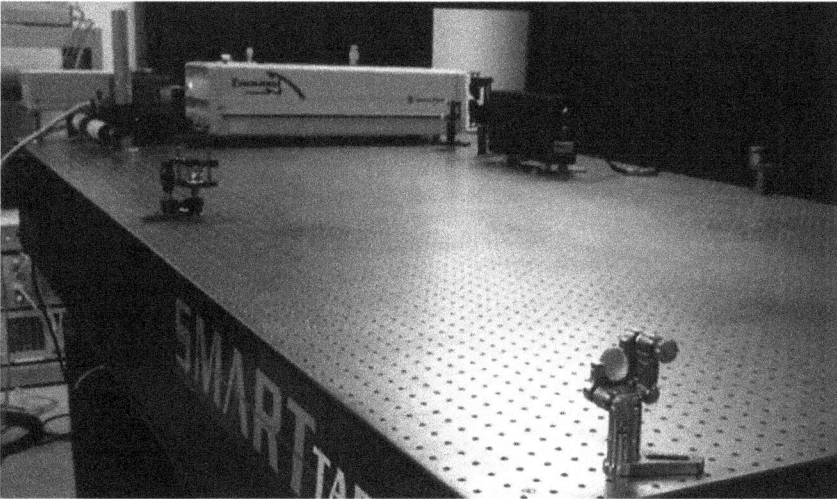

**Figure 1.4.** Michelson interferometer assembled on an optical table.

22 Hz, the effect of vibration on the image is an observable jitter or "jiggle". At higher frequencies, it is perceived as blurring (see [4]).

The second example is a Michelson interferometer assembled on an optical table (Fig. 1.4). The image is projected, through a lens, to a screen placed just off the table to the left, not seen in the photograph. For sake of experiment, the table was subjected to random vibration excitation. Figure 1.5 compares interferometer images taken with

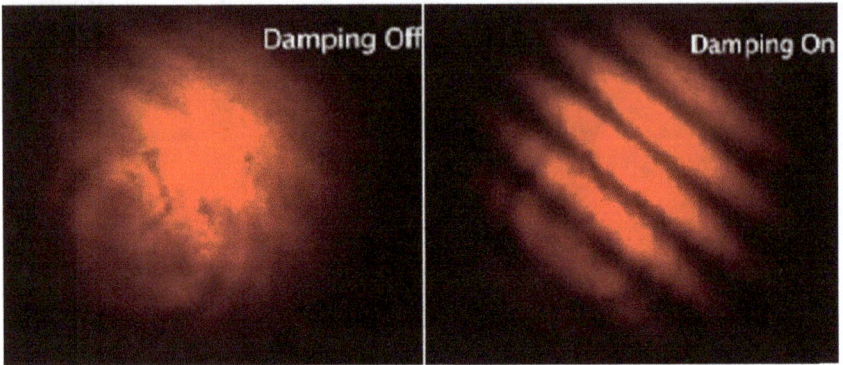

**Figure 1.5.** Interferometer images taken on the platform protected from vibration (right) and not protected from vibration (left). Courtesy of Newport Corporation. All rights reserved.

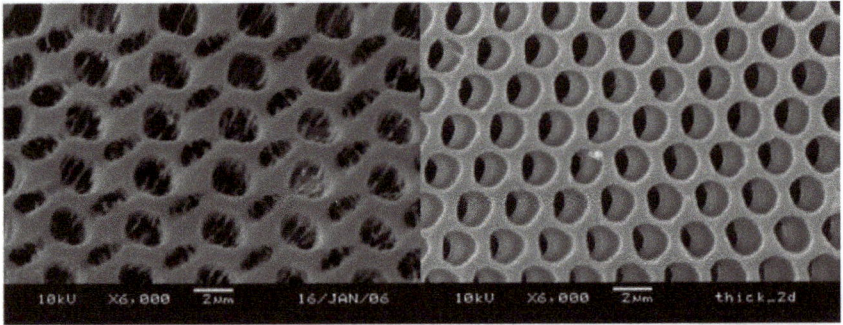

**Figure 1.6.** Nanostructures created on the platform protected from vibration (right) and not protected from vibration (left) [5]. Permission to use granted by Newport Corporation. All rights reserved.

resonance vibration of the table suppressed and not suppressed. In this case the vibration, mostly in the 200 Hz range, makes the fringes completely blurry. Resonance properties of optical tables and relevant methods of vibration control will be considered in Chapters 6, 7 and 9.

Another example, provided by Dr. Steven Kooi [5], concerns a nanostructure built on a vibration-isolated optical table. The images in Fig. 1.6 show the results of a multi-beam interference lithography setup. The images come from a four-beam exposure. One beam from a continuous wave 532 nm diode laser is split into four beams.

The operator controls the relative intensity, polarization, and angle of intersection of the four beams that are recombined at the sample. The sample is a photoresist material. The 3D interference pattern is transferred into photoresist during exposure and, after developing, a 3D structured polymer is created. The exposure times can be relatively long (minutes), so any vibration during exposure can cause degradation in the structure. The picture on the left shows a structure severely influenced by vibration that created a wave pattern of defects. Vibration sensitivity is due to the fact that the application is dependent on exact alignment of the beams and therefore sensitive to flexural vibration of the table. The defects are not seen in the picture on the right showing the result with the active (electronic) dampers enabled. This vibration control technique is explained in detail in Chapter 9. In this example vibration control made all the difference for the application.

The next example, contributed by Nicholas Geisse and Kevin Kit Parker [5], illustrates the influence of vibration on another microscopy application, this time on the atomic force microscopy of cardiac tissue. Figure 1.7 shows the force curve describing the deflection of the cantilever probe vs. the displacement of the piezo motor in presence of vibration and with vibration suppressed by active damping. The force curves had been taken on the same cell, at the same spot within one minute of each other. The influence of vibration and the effect of vibration control are quite evident from these graphs.

The last example of this chapter, contributed by Dr. Tommaso Baldacchini based on the joint experiment with the author, illustrates vibration sensitivity of images obtained by portable atomic force microscope TraxAFM from Nanoscience Instruments. Figure 1.8 compares scans of a 100 nm trench in a silicon wafer and a 4 $\mu$m × 4 $\mu$m piece of a Blue-ray DVD obtained by imaging on an active vibration isolation workstation (see Section 9.1 of Chapter 9) with active vibration control switched on and off. The comparison clearly shows the imaging noise introduced by vibration and the effect of active vibration isolation.

In summary, vibrations affecting optomechanical performance vary by source (external, internal, and on-board), by character (stationary vs. transient; forced vs. resonant), and frequency content. Different applications may be sensitive to different vibrational inputs. Accordingly, vibration control methods and means applicable in

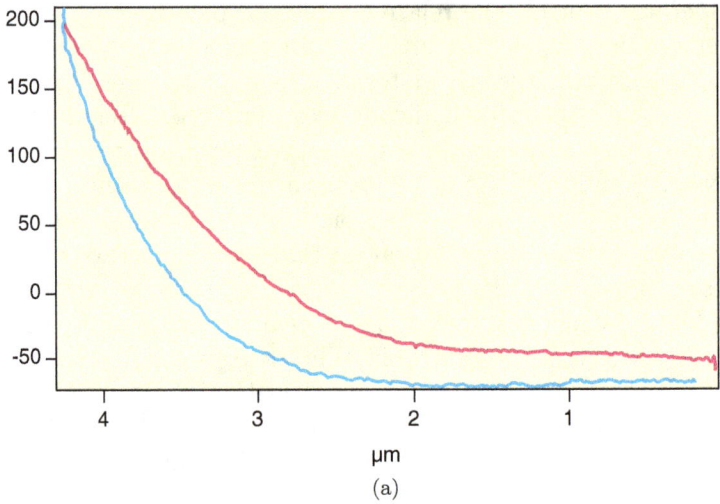

(a)

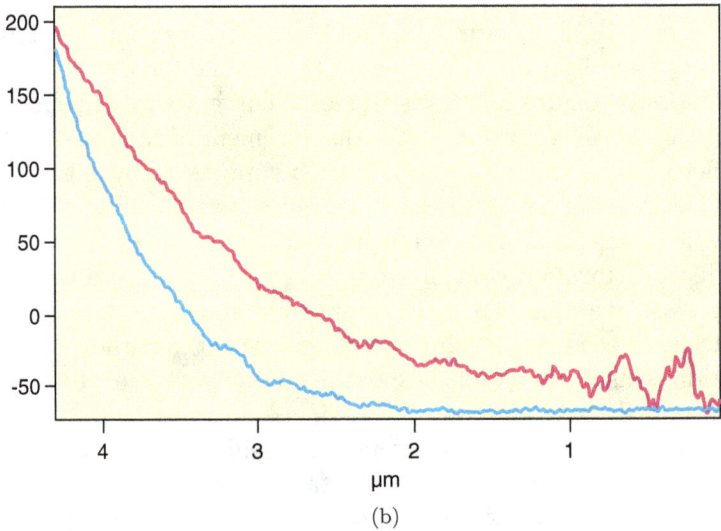

(b)

**Figure 1.7.** Force curves obtained during atomic force microscopy of cardiac tissue. (a) Vibration suppressed; (b) vibration not suppressed [5]. Permission to use granted by Newport Corporation. All rights reserved.

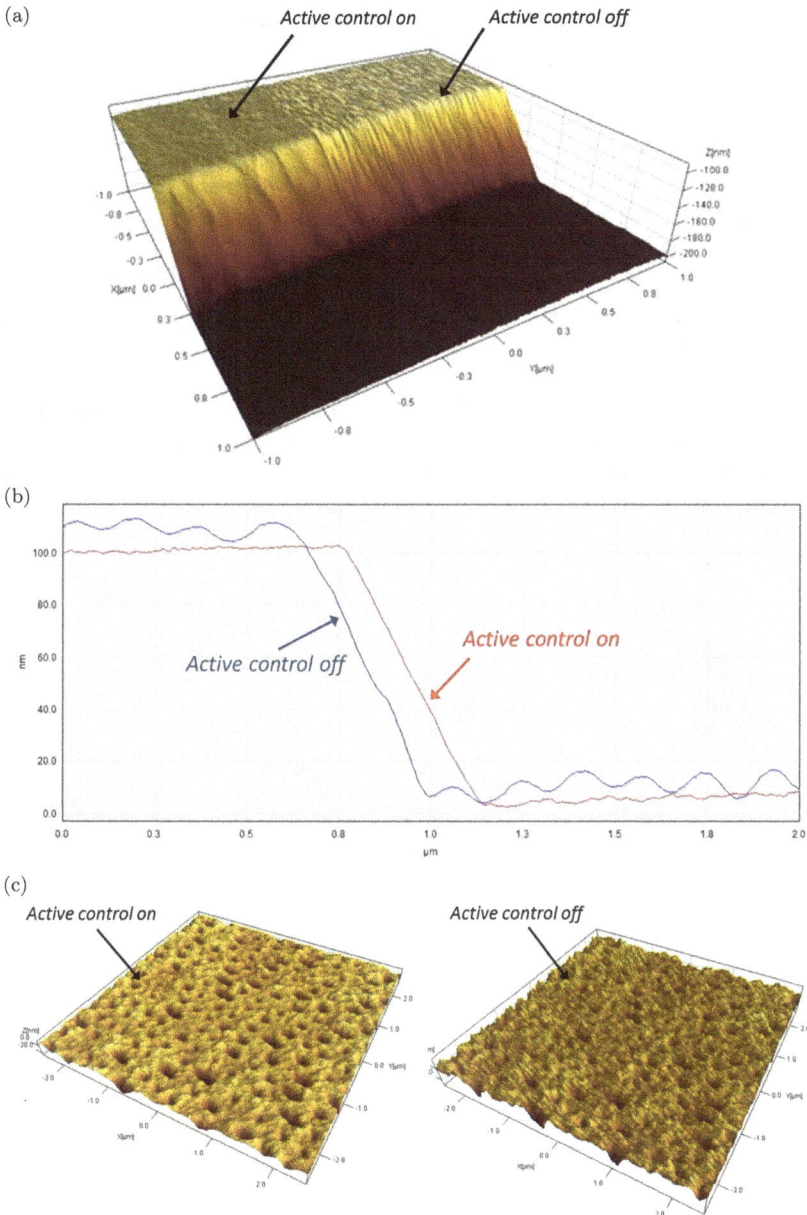

**Figure 1.8.** AFM images of a 100 nm trench in a silicon wafer (a) and a 4 $\mu$m $\times$ 4 $\mu$m piece of a Blue-ray DVD (b). Courtesy of Newport Corporation. All rights reserved.

every situation may vary, too. This will be discussed in subsequent chapters.

## References

1. ANSI/ASA S1.1, *American National Standard Acoustical Terminology*, Acoustical Society of America, Melville, NY, 2013, 81 p.
2. K.B. Doyle, V.L. Genberg, and G.J. Michels, *Integrated Optomechanical Analysis*, 2nd edn., SPIE Press, Bellingham, WA, 2012, 408 p.
3. https://www.youtube.com/watch?v=j3kB9PN7lYI, 2:32–3:12. Accessed on March 1, 2021.
4. H. Amick and M. Stead, Vibration sensitivity of laboratory bench microscopes, *Sound and Vibration*, February 2007, p. 10–16.
5. *Newport Resource*, Catalog, Newport Corporation, 2008 and later editions.

Chapter 2

# Mathematical Methods of Vibration Studies

Mathematical notions and methods pervade vibration studies. Mathematics is the language Nature speaks to us. An engineer is much better off understanding this language than trying to avoid it. The circle of mathematical facts and methods useful for vibration control is very wide, well worthy of a separate book. The goal of this chapter is to refresh basic mathematics necessary for adequate description of mechanical models and signals encountered in vibration control. We'll concentrate on frequency methods based on Fourier analysis and its applications.

## 2.1　Harmonic Vibration

To review the ways of describing vibration, we start from the simplest form of vibration that has fundamental importance, although it seldom occurs in nature and technical applications in its pure form. It is harmonic, or sinusoidal, vibration. It is characterized by the *period*, $T$, and the *amplitude*, or *peak value*, $A_p$ (see Fig. 2.1). The *frequency*, $f$, and the *circular frequency*, $\omega$, are defined in terms of the period:

$$f = \frac{1}{T}, \quad \omega = 2\pi f.$$

During harmonic vibration, a mechanical parameter changes in time following a sinusoidal function of time. The *phase*, $\varphi$, is introduced

11

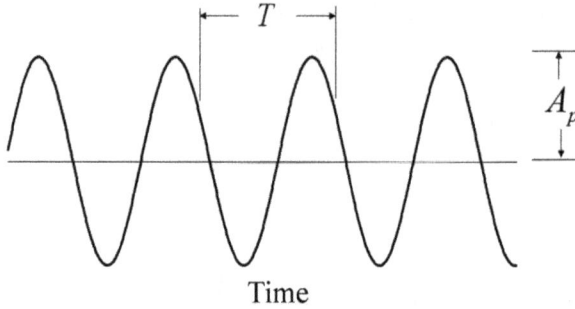

**Figure 2.1.**    Sinusoidal vibration.

to describe the shift of the oscillation in time relative to another harmonic signal or a fixed origin on the time axis. So, the sinusoidal vibration depicted in Fig. 2.1 can be described by the following equation:

$$A(t) = A_p \cos(\varphi + \omega t).$$

The following parameters are also used to characterize the scope of the sinusoidal vibration: *peak-to-peak* value:

$$A_{pp} = 2A_p,$$

*mean absolute* value,

$$A_{\text{av}} = \frac{1}{T} \int_{\tau}^{\tau+T} |A(t)|\, dt = \frac{2}{\pi} A_p \approx 0.637\, A_p,$$

and *root mean square* (RMS) value,

$$A_{\text{rms}} = \sqrt{\frac{1}{T} \int_{\tau}^{\tau+T} [A(t)]^2 dt} = \frac{1}{\sqrt{2}} A_p \approx 0.707\, A_p.$$

As noted before, the three parameters describing mechanical vibration — displacement, velocity, and acceleration — are interrelated. Their relationships are especially simple in case of harmonic vibration. If the displacement is represented as

$$u(t) = u_p \cos(\phi_0 + \omega t), \qquad (2.1)$$

then, by definition, velocity is

$$v(t) = \frac{du(t)}{dt} = -\omega u_p \sin(\phi_0 + \omega t) = \omega u_p \cos(\phi_0 + \tfrac{\pi}{2} + \omega t),$$

and acceleration is

$$w(t) = -\omega^2 u_p \cos(\phi_0 + \omega t) = \omega^2 u_p \cos(\phi_0 + \pi + \omega t).$$

So, velocity has $(\pi/2)$, or $90°$, phase shift and $\omega = 2\pi f$ amplitude factor with respect to the displacement; acceleration has $\pi$, or $180°$, phase shift and $\omega^2 = 4\pi^2 f^2$ amplitude factor with respect to the displacement.

**Exercise 2.1.** (a) An air compressor rotates at 3000 revolutions/min. The misbalance creates a vibrational force. What is the frequency of this force in Hz? What is the circular frequency? What is the period of vibration?

**Answer:** 50 Hz; 314 rad/s; 0.02 s.

(b) A pendulum oscillates with the period of 0.5 s and amplitude of 3 mm. What are the RMS values of displacement, velocity, and acceleration?

**Answer:** 2.12 mm; 26.7 mm/s; 335 mm/s$^2$, or 34.1 milli-g.

## 2.2 Complex Amplitudes

Complex amplitude is a very useful notion that simplifies operations with vibrational quantities. A complex number is defined by two real numbers: a real part $x$ and an imaginary part $y$,

$$z = x + iy,$$

where $i$ is the so-called imaginary unit, or the square root of $-1$,

$$i^2 = -1.$$

A complex number can be written also in polar form using its modulus, $|z|$, and argument, $\varphi$:

$$z = |z|e^{i\varphi},$$

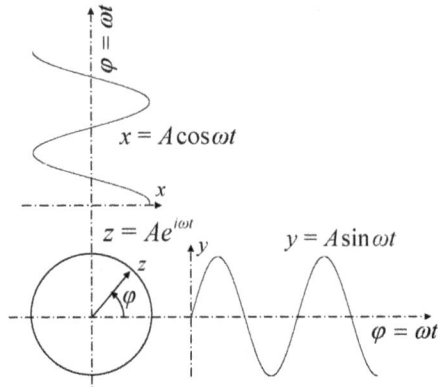

**Figure 2.2.** Complex number in polar form, and the relationship between rotational and sinusoidal motions.

where, by definition,

$$|z| = \sqrt{x^2 + y^2}, \quad e^{i\varphi} = \cos\varphi + i\sin\varphi.$$

It is instructive to consider a complex number as a vector on the complex plane. If such a vector, with a given length, or modulus, is rotating uniformly about the origin, its projections generate sinusoidal functions that describe harmonic vibration (see Fig. 2.2).

A harmonic vibration with peak value, or real amplitude, $u_p$, as described by Eq. (2.1), can be represented as a real part of a complex function using a complex exponent that includes an initial phase $\varphi_0$:

$$u(t) = u_p \cos(\varphi_0 + \omega t) = \text{Re}\left[u_p e^{i(\varphi_0 + \omega t)}\right].$$

This can be re-written as

$$u(t) = \text{Re}\left(u_c e^{i\omega t}\right),$$

where

$$u_c = u_p e^{i\varphi_0}.$$

The complex number $u_c$ is called the complex amplitude. It contains information about both amplitude and phase of the harmonic vibration. Real amplitude equals the modulus of the complex amplitude: $u_p = |u_c|$.

Complex amplitudes are convenient because they reduce all linear transformations of multiple harmonic signals with the same frequency to linear algebraic operations. Suppose we have two harmonic signals with the same amplitude but different phases, like two dynamic forces caused by the same rotating misbalance but arriving through different structural paths:

$$u_1(t) = a_1 \cos(\varphi_1 + \omega t); \quad u_2(t) = a_2 \cos(\varphi_2 + \omega t).$$

Complex amplitudes are $u_{1c} = a_1 e^{i\varphi_1}$ and $u_{2c} = a_2 e^{i\varphi_2}$. Then the sum of the two signals is

$$u_1(t) + u_2(t) = \mathrm{Re}[(u_{1c} + u_{2c})e^{i\omega t}].$$

This means that the complex amplitude of the sum of two signals is the sum of their complex amplitudes. The real amplitude of the summary signal is the absolute value of the sum of complex amplitudes:

$$\max[u_1(t) + u_2(t)] = |u_{1c} + u_{2c}|.$$

**Exercise 2.2.** Two forces from rotating equipment are applied to the common foundation. The first one has amplitude 1 N; the second one has amplitude 2 N and its phase is shifted 60° with respect to the first one. Find the amplitude of the resulting force using complex amplitudes and polar representation of complex numbers.

**Solution:** $u_{1c} = 1$; $u_{2c} = 2e^{i\frac{\pi}{3}} = 2\left(\frac{1}{2} + i\frac{\sqrt{3}}{2}\right)$; $u_{1c} + u_{2c} = 2 + i\sqrt{3}$. $|u_{1c} + u_{2c}| = \sqrt{7} \simeq 2.646\,\mathrm{N}$. See also the graphic solution in Fig. 2.3.

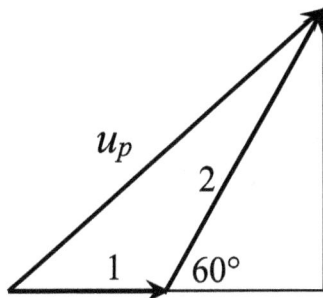

**Figure 2.3.** Illustration to Exercise 2.2.

Relationships between complex amplitudes of displacement, velocity, and acceleration of the harmonic vibration reflect both amplitude and phase relations: since

$$\frac{d}{dt}(e^{i\omega t}) = i\omega e^{i\omega t},$$

complex amplitudes of displacement, velocity, and acceleration are, respectively,

$$u_c = u_p e^{i\varphi_0}, v_c = i\omega u_c, w_c = i\omega v_c = -\omega^2 u_c.$$

So, linear operations on harmonic values with the same frequency, including summation, scaling, differentiation, and integration, can be performed algebraically using complex amplitudes. Differential equations become algebraic equations in complex amplitudes. Complex representation of harmonic values is widely used in analyzing mechanical models. We'll apply this technique in the next chapter to basic models of vibrating systems and their transfer functions.

## 2.3   Fourier Series and Fourier Transforms

Spectrum is the most important notion that can be used for describing the structure of a vibration signal. By the ANSI/ASA definition [1], a spectrum is "a description, for a function of time, of the resolution of a signal into components, each of different frequency and, usually, different amplitude and phase." This wide definition covers any representation of a signal by a collection of frequencies and corresponding complex amplitudes. Three kinds of spectra are used most often in vibration analysis: linear spectrum, power spectral density (PSD), and $\frac{1}{3}$ octave spectrum. We'll consider all three in some detail in subsequent sections.

The most well-known spectral representation of a time function is the *Fourier series* that is valid for periodic functions. Any continuous periodic function can be represented by an infinite sum of sine and cosine functions:

$$x(t) = \frac{a_0}{2} + \sum_{n=1}^{\infty} (a_n \cos n\omega t + b_n \sin n\omega t). \tag{2.2}$$

Fourier coefficients $a_n$ and $b_n$ are given by the formulas

$$a_n = \frac{2}{T} \int_0^T x(t) \cos n\omega t \, dt, \quad b_n = \frac{2}{T} \int_0^T x(t) \sin n\omega t \, dt,$$

where $T$ is the period, $\omega = 2\pi/T$. Detailed theory of Fourier series can be found in many excellent textbooks [2,3].

In mechanical terms, $t$ is time, and Eq. (2.2) represents periodic vibration by a sum of harmonic components. The next formula shows these harmonic vibration components re-written in agreement with the definition (2.1) as cosine functions of time with certain amplitudes and phases:

$$x(t) = \frac{a_0}{2} + \sum_{n=1}^{\infty} \sqrt{a_n^2 + b_n^2} \, \cos(n\omega t - \phi_n), \quad \phi_n = \tan^{-1}\left(\frac{b_n}{a_n}\right).$$

The same real signal can be presented using complex amplitudes of spectral components:

$$x(t) = \sum_{n=-\infty}^{\infty} c_n e^{in\omega t} = c_0 + \operatorname{Re}\left\{\sum_{n=1}^{\infty} 2c_n e^{in\omega t}\right\},$$

where the complex Fourier coefficients are given by

$$c_n = \frac{1}{T} \int_0^T x(t) e^{-in\omega t} dt.$$

Complex Fourier coefficients are related to real Fourier coefficients by the following equations:

$$c_{-n} = \tfrac{1}{2}(a_n + ib_n), \quad c_n = \tfrac{1}{2}(a_n - ib_n).$$

An important property of Fourier series is expressed by Parseval's formula:

$$\left\langle x^2(t) \right\rangle = \frac{1}{T} \int_0^T x^2(t) dt = \sum_{n=-\infty}^{\infty} |c_n|^2$$

$$= c_0^2 + \sum_{n=1}^{\infty} 2|c_n|^2 = \left(\frac{a_0}{2}\right)^2 + \sum_{n=1}^{\infty} \tfrac{1}{2}(a_n^2 + b_n^2).$$

(2.3)

According to this equation, the average square of any periodic value equals sum of the squares of spectral amplitudes. In physical language, it means that the total energy of the signal equals sum total of the energies of its spectral components.

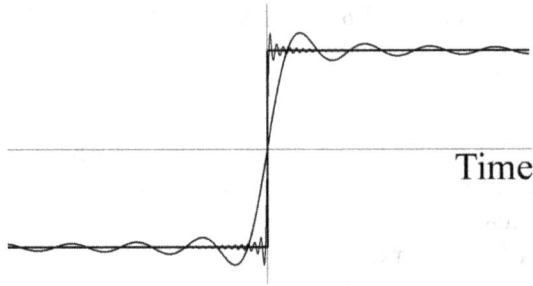

**Figure 2.4.** Gibbs effect illustrated by Fourier approximations of the step function retaining 10 and 100 terms.

If the periodic function is not continuous, the corresponding Fourier series can be constructed, but it does not converge uniformly. This case is important for vibration control because vibration signals are often analyzed by repeating the data from a standard interval periodically, which may result in discontinuities at the ends of the interval. Applying Fourier series in this situation gives rise to so-called Gibbs effect: summation of the series produces a ripple, or a shock effect, which does not go away even as the number of Fourier components tends to infinity. This is illustrated in Fig. 2.4 with graphs of Fourier approximations of the step function retaining 10 and 100 terms. The value of the overshoot is, approximately, 0.089 of the value of discontinuity [3]. Special methods (windowing) are applied in signal processing to avoid the Gibbs effect.

*Fourier transform* expands the idea of Fourier series to non-periodic functions. A non-periodic function can be treated as a periodic function with an infinite period. In the limit as period tends to infinity, summation can be replaced by integration. The detailed mathematical derivation [2, 3] is beyond the scope of this book; the result is presented in Eq. (2.4):

$$x(t) = \tfrac{1}{2\pi} \int_{-\infty}^{\infty} X(\omega)e^{i\omega t}d\omega. \tag{2.4}$$

The function $X(\omega)$ is called a Fourier image of the signal $x(t)$. The inverse Fourier transform is given by

$$X(\omega) = \int_{-\infty}^{\infty} x(t)e^{-i\omega t}dt.$$

Again, an energy type equality, an analog to Parseval's formula, connects the integrals of signal and its transform squared in time and frequency domains. It is called Plancherel's theorem. The equation is

$$\int_{-\infty}^{\infty} |x(t)|^2 dt = \int_{-\infty}^{\infty} |X(\omega)|^2 \frac{d\omega}{2\pi}.$$

Fourier transform theory forms the basis for the discrete Fourier transform (DFT) that is an important instrument of experimental vibration analysis considered in Chapter 4.

## 2.4   Random Vibration

Application of spectral analysis to vibration signals depends on the kind of vibration we are dealing with: deterministic or random. In simple intuitive terms, deterministic vibration is the one that can be predicted at any future time from past knowledge. Random vibration cannot be predicted from past knowledge except in some average, or statistical, sense. Returning to what was said in the previous chapter about typical sources of vibration in buildings, machinery usually creates deterministic vibration; background activities such as footfall or wind load create random excitation.

The notion of random vibration is illustrated in Fig. 2.5. It shows three one-second snippets of vibration on the top surface of the author's desk recorded by an accelerometer (see Section 4.1 of Chapter 4) and a signal acquisition system (see Section 4.2 of Chapter 4). This is what random vibration looks like. The time histories are all different, but from their appearance you can easily believe that they are related, belonging to the same process or phenomenon.

Vibration under uncertain disturbances represents a random process [4]. Each outcome under each occasion is called a *sample function*. A set of sample functions is called an *ensemble*. So, Fig. 2.5 shows an ensemble.

The main tool for studying random vibrations is averaging. The ergodic hypothesis [4] states that averaging over an ensemble can be replaced by averaging in the time domain. It is applicable to stationary processes, that is, the processes with statistical characteristics that are not changing in time. For stationary processes defined in

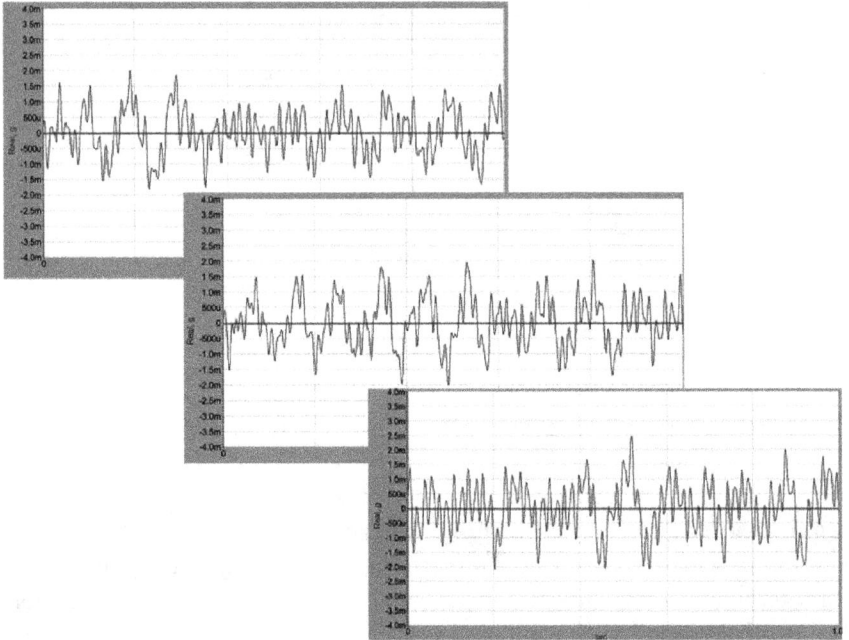

**Figure 2.5.** Random vibration (acceleration vs. time).

the narrow sense, these characteristics include mean square,

$$\langle x^2 \rangle \cong \frac{1}{T} \int_0^T x^2(t)dt, \qquad (2.5)$$

and autocorrelation function,

$$R_{xx}(\tau) \cong \frac{1}{T} \int_0^T x(t)x(t+\tau)dt. \qquad (2.6)$$

In Eqs. (2.5) and (2.6), $T$ is assumed to be a sufficiently long integration period. To compare two processes, the cross-correlation function is used, given by

$$R_{xy}(\tau) \cong \frac{1}{T} \int_0^T x(t)y(t+\tau)dt. \qquad (2.7)$$

For stationary processes, functions (2.5)–(2.7) do not depend on the starting point in time.

## 2.5 Power Spectral Density and One-third Octave Spectrum

Frequency composition of random vibration is often described by its PSD. By definition [1], PSD $W_x(f)$ of a random variable $x(t)$ equals the mean square response of an ideal narrow-band filter to $x(t)$ divided by the bandwidth $\Delta f$ in the limit as $\Delta f \to 0$ at frequency $f$ (Hz):

$$W_x(f) = \lim_{\Delta f \to 0} \frac{\langle x_{\Delta f}^2 \rangle}{\Delta f}. \tag{2.8}$$

Alternatively, PSD can be defined formally through Fourier transform of the autocorrelation function (2.6). Since the autocorrelation function is an even function of frequency, a special form of Fourier transform is used, called the Fourier cosine transform [4]:

$$W_x(f) = 4 \int_0^\infty R_{xx}(\tau) \cos(2\pi f \tau) d\tau.$$

So, PSD is a cosine Fourier image of the auto-correlation function.

Note the relation between the linear spectrum and the PSD: according to (2.3), the component of the linear (RMS) spectrum at the frequency $f_n$ equals

$$G_{\mathrm{RMS}x}(f_n) = \sqrt{\frac{1}{2}(a_n^2 + b_n^2)}.$$

On the other hand, using (2.3) in the definition (2.8) gives

$$W_x(f_n) = \frac{1}{\Delta f} \frac{1}{2}(a_n^2 + b_n^2) = \frac{1}{\Delta f}(G_{\mathrm{RMS}x}(f))^2,$$

or

$$(G_{\mathrm{RMS}x}(f)) = \sqrt{W_x(f_n)}\sqrt{\Delta f}.$$

This means that the PSD is the RMS linear spectrum squared, divided by the frequency bin of the analysis. Accordingly, the RMS linear spectrum equals square root of the product of PSD and the frequency bin. For random vibration with continuous spectrum, PSD does not depend on $\Delta f$, provided the latter is sufficiently small.

On the other hand, linear spectrum depends on the size of the frequency bin of the analysis. Therefore, vibration survey results expressed as linear spectra must always be accompanied by an indication of the frequency resolution.

Power spectral densities play an important role in signal processing and analysis of linear systems. Mean square of a signal can be calculated from its PSD by integration:

$$\langle x^2 \rangle = \int_0^\infty W_x(f)df. \tag{2.9}$$

A linear relationship between two vibration signals can be described in the frequency domain by a transfer function $H(f)$, so that, at any frequency $f$, complex amplitude of the output signal, $y(f)$, is related to the complex amplitude of the input signal, $x(f)$, by the linear relationship

$$y(f) = H(f)x(f).$$

(Multiple examples of transfer functions derived from mechanical models of vibrating systems are presented in the next chapter.) In this case, the power spectral densities of input and output signals are related by

$$W_y(f) = |H(f)|^2 W_x(f). \tag{2.10}$$

The unit of measurement of the PSD is the engineering unit squared divided by Hertz. For example, the PSD of acceleration can be measured in $g^2/\text{Hz}$. Often the square root of PSD is used, measured in units such as $g/\sqrt{Hz}$. These units are not intuitive, and their relation to the real-world vibration magnitudes is not transparent.

To have the spectra measured in engineering units but independent on the frequency bin of the analysis, constant percentage spectra are employed. Octave spectra and one-third octave spectra are examples of constant percentage spectra. One-third octave spectra gained wide acceptance in describing random vibrations.

Two frequencies $f_1$ and $f_2$ are divided by an octave if $f_2/f_1 = 2$. Two frequencies $f_1$ and $f_2$ are divided by one-third of an octave if $f_2/f_1 = 2^{1/3} \approx 1.2599$. One-third octave frequency bands are stipulated by the standard ANSI S1.11-2004. According to this standard,

**Table 2.1.** Standard 1/3 octave frequency bands.

| Band Number | Nominal Centre Frequency (Hz) | Passband (Hz) | Band Number | Nominal Centre Frequency (Hz) | Passband (Hz) |
|---|---|---|---|---|---|
| 1 | 1.25 | 1.12–1.41 | 11 | 12.5 | 11.2–14.1 |
| 2 | 1.6 | 1.41–1.78 | 12 | 16 | 14.1–17.8 |
| 3 | 2 | 1.78–2.24 | 13 | 20 | 17.8–22.4 |
| 4 | 2.5 | 2.24–2.82 | 14 | 25 | 22.4–28.2 |
| 5 | 3.15 | 2.82–3.55 | 15 | 31.5 | 28.2–35.5 |
| 6 | 4 | 3.55–4.47 | 16 | 40 | 35.5–44.7 |
| 7 | 5 | 4.47–5.62 | 17 | 50 | 44.7–56.2 |
| 8 | 6.3 | 5.62–7.08 | 18 | 63 | 56.2–70.8 |
| 9 | 8 | 7.08–8.91 | 19 | 80 | 70.8–89.1 |
| 10 | 10 | 8.91–11.2 | 20 | 100 | 89.1–112 |

the frequency bands are adjusted so that the center of the 30th band equals 1000 Hz, and somewhat rounded. Centers and passbands of the first 20 standard 1/3 octave frequency bands are given in Table 2.1. Centers, lower and upper frequencies for higher 1/3 octave bands can be found in the textbook [5]. One-third octave center frequencies are placed uniformly on the logarithmic frequency scale.

A continuous PSD $W(f)$ can be recalculated into a 1/3 octave spectrum by integration over the 1/3 octave frequency bands $(f_k^-, f_k^+)$:

$$(S_{1/3})_k = \sqrt{\int_{f_k^-}^{f_k^+} W(f)df}. \qquad (2.11)$$

However, PSD and narrow-band linear spectra cannot be restored from the 1/3 octave spectrum, unless additional *a priori* assumptions are made.

One-third octave spectra are measured in engineering units; they can give some notion of the actual levels of vibration. In particular, for the bands dominated by tonal vibrations, the one-third octave measurement would be only slightly higher than the tonal amplitude; for the bands containing several tones and broadband noise, the one-third octave component will be higher than the individual

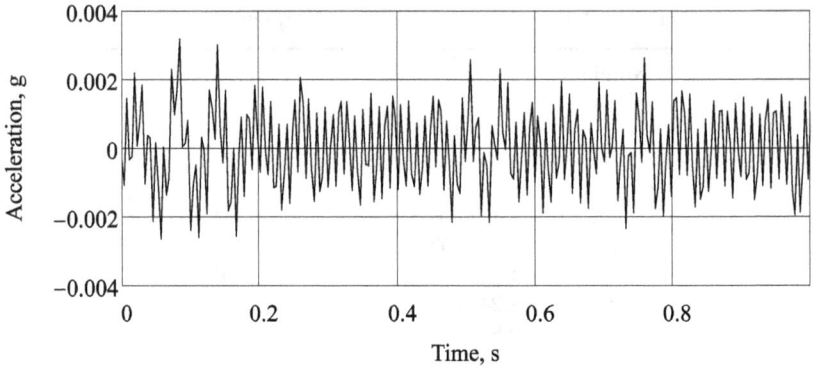

**Figure 2.6.**   Acceleration signal: fragment of time history.

tone amplitudes, reflecting their cumulative effect. Amick and Bui [6] give a useful discussion of the relationship between different types of spectra. Many experts consider one-third octave spectra containing "just enough" details about real-life vibrations to estimate the severity of vibration environments; one-third octave RMS spectra of velocity form the basis of vibration criteria in form of "VC levels" widely used for estimating whether certain vibration environments are suitable for particular applications. This will be discussed in more detail in Chapter 5.

**Example 2.1. Various spectral representations of the same vibration signal:** For sake of illustration, consider the same vibration signal represented in different ways. Acceleration was measured by a piezoelectric accelerometer and sampled at the frequency $f_s = 256$ Hz by a signal analyzer (see Chapter 4). Figure 2.6 shows a one-second snippet of the time history. Figure 2.7 shows linear spectra with $\Delta f = 0.25$ Hz resolution of acceleration and velocity, as well as PSD of acceleration plotted in RMS units. The acceleration spectrum shows higher frequency components more prominently, compared to the velocity spectrum. The square root of the PSD spectrum is exactly proportional to the linear spectrum with the coefficient depending on $\Delta f$. Figure 2.8 shows one-third octave spectrum of RMS velocity as required by the VC criteria discussed in Chapter 5.

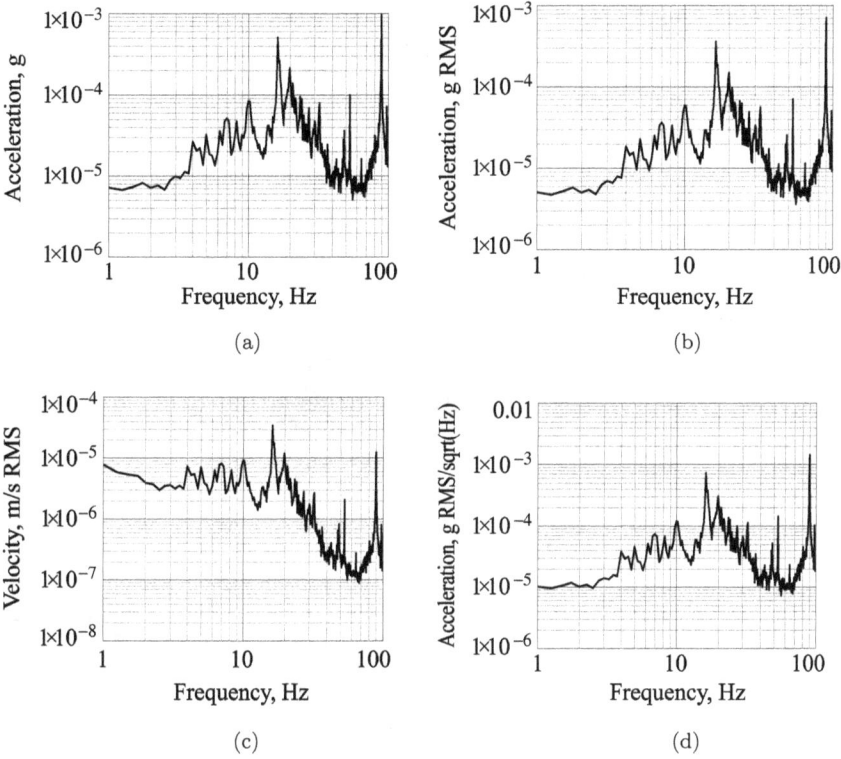

**Figure 2.7.** Different formats of the constant narrow band (0.25 Hz) spectrum of the same vibration signal. (a) Acceleration amplitude; (b) acceleration RMS; (c) velocity RMS; (d) acceleration, square root of PSD.

## 2.6 Decibel Scale

Use of decibel scale for measuring physical quantities originates in acoustics and is rooted in the fact that human perception of loudness follows logarithmic rather than linear scale of sound pressure or acoustical power. Decibel scale is useful in vibration studies because oftentimes dealing with vibration signals and transfer functions involves comparing values that differ in orders of magnitude.

Decibel is a measure of a quantity relative to some reference quantity of the same physical nature. Specifically, if we consider a mechanical quantity, say, mechanical displacement $u$, and a reference displacement, $u_{ref}$, had been established, then the measure of

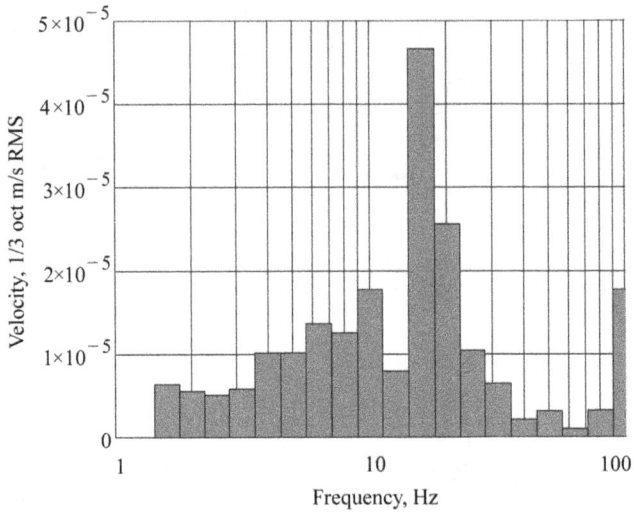

**Figure 2.8.**   One-third octave velocity spectrum of the same signal.

$u$ in decibels, $L_u$, would be 20 times the common logarithm of their ratio:

$$L_u = 20 \log_{10} \frac{|u|}{|u_{\text{ref}}|}.$$

Unfortunately, there are no universally accepted reference levels for mechanical displacements, velocities, or accelerations, so when you see measurement data of mechanical quantities in decibels, you should always ask: related to what reference?

Decibels are widely used to characterize vibration transmissibility through mechanical systems, which involves comparing the quantities of the same kind (see Chapter 3 for examples). If the transmissibility is described by a non-dimensional function of frequency, $T(f)$, then this transfer function in decibels is 20 times the common logarithm of its absolute value:

$$L = 20 \log_{10} |T(f)|.$$

Table 2.2 shows the relationship between the ratio and the decibel measure. For example, if the effectiveness of a certain vibration control device is 20 dB, this means that it reduces vibration by the factor of 0.1, or 10 times.

**Table 2.2.** Comparison of linear scale and decibel scale.

| Ratio | 0.01 | 0.03 | 0.05 | 0.1 | 0.25 | 0.3 | 0.5 | 1 |
|---|---|---|---|---|---|---|---|---|
| dB | −40 | −30.5 | −26.0 | −20 | −12.0 | −10.5 | −6.0 | 0 |
| Ratio | 2 | 3 | 4 | 10 | 20 | 30 | 100 | 1000 |
| dB | 6.0 | 9.5 | 12.0 | 20 | 26.0 | 29.5 | 40 | 60 |

**Exercise 2.3.** Vibration isolator reduces vibration on the isolated platform in the 13th third-octave band by 30 dB; in the 10th third-octave band by 18 dB compared with that on the floor. What is the ratio of vibration amplitudes on the floor and on the platform: (a) at 10 Hz? (b) at 20 Hz?

**Answer:** (a) 7.9; (b) 31.6.

## References

1. ANSI/ASA S1.1, *American National Standard Acoustical Terminology*, Acoustical Society of America, Melville, NY, 2013, 81 p.
2. H. Dym and H.P. McKean, *Fourier Series and Integrals*, Academic Press, New York, 1985, 295 p.
3. D.L. Kreider, R.G. Kuller, D.R. Ostberg, and F.W. Perkins, *An Introduction to Linear Analysis*, Addison-Wesley, Reading, Massachusetts, 1966, ix, 773 p.
4. J.S. Bendat and A.G. Piersol, *Measurement and Analysis of Random Data*, John Wiley & Sons, New York, 1966, 390 p.
5. A.D. Pierce, *Acoustics: An Introduction to Physical Principles and Applications*, Acoustical Society of America, Woodbury, NY, 1991, p. 58.
6. H. Amick and S.K. Bui, A review of several methods for processing vibration data, *Proceedings of International Society for Optical Engineering (SPIE)*, *Vol. 1619*, San Jose, CA, November 4–6, 1991, pp. 253–264.

Chapter 3

# Basic Models of Mechanical Oscillatory Systems

This chapter reviews mechanical models of vibrating mechanical systems. In this study, we'll make use of the mathematical methods discussed in Chapter 2. The models that are reviewed here are universal. They give insights into properties of devices used for vibration control, such as vibration isolators; on the other hand, they can be considered as models of vibration measurement devices, such as accelerometers, and from yet another perspective they can be applied to the very objects of vibration control, namely optical and optomechanical systems. When pondering vibrational behavior of a physical object, an engineer usually thinks in terms of a mechanical model. These models will be used extensively in subsequent chapters.

In order of increased complexity, there are undamped and damped systems; single-degree-of-freedom (SDOF), multi-degree-of-freedom (MDOF), and continuous systems. Two specific models of practical importance — an elastically supported rigid body, modeling a vibration-isolated platform, and a beam carrying an inertial payload, modeling an optomechanical post — will be considered in some detail; finally, a simple model of a flexible vibration-isolated body will be analyzed. We begin with SDOF systems.

## 3.1 Simple Linear Oscillator without Damping

The simplest model of an oscillatory system in vibration control, as well as in other technical disciplines, is a linear oscillator. It includes

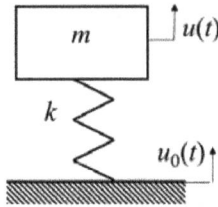

**Figure 3.1.**　SDOF linear oscillator without damping.

a concentrated mass, $m$, which is connected by a spring, $k$, to a rigid foundation (see Fig. 3.1). Assume for a moment that the foundation moves, and its displacement is described by the function $u_0(t)$. The mechanical system responds with the motion of the mass $m$ described by $u(t)$. This represents the basic idea of vibration isolation, where the foundation is the source of vibration, say, vibrating floor, and the mass $m$ depicts the object being protected.

To find out how the motion of the floor is transmitted to the mass $m$, we need to write the equation of motion. It is given by Newton's second law: force equals mass time acceleration, which is by definition the second derivative of displacement with respect to time; the force is in our case the compression force of the spring, expressed by Hooke's law:

$$m\ddot{u} = F_s, \quad F_s = -k(u - u_0), \tag{3.1}$$

so that

$$m\ddot{u}(t) + ku(t) = ku_0(t). \tag{3.2}$$

This is a differential equation for displacement. Mass and stiffness are parameters in the equation. Together, they define the most important characteristics of the oscillator — its natural frequency. Specifically,

$$\omega_0 = \sqrt{\frac{k}{m}}, \tag{3.3}$$

is the circular natural frequency, and

$$f_0 = \frac{1}{2\pi}\sqrt{\frac{k}{m}}, \tag{3.4}$$

is the natural frequency in Hz. Note that the natural frequency increases with the increase of stiffness and decreases with the increase of mass.

In terms of the natural frequency, Eq. (3.2) takes the form

$$\ddot{u}(t) + \omega_0^2 u(t) = \omega_0^2 u_0(t). \tag{3.5}$$

This equation requires two initial conditions, for example, displacement and velocity at the initial moment. With these initial conditions, the exact solution of Eq. (3.5) is

$$u(t) = u(0)\cos\omega_0 t + \frac{\dot{u}(0)}{\omega_0}\sin\omega_0 t + \omega_0 \int_0^t u_0(\tau)\sin\omega_0(t-\tau)d\tau. \tag{3.6}$$

The right-hand side of Eq. (3.6) consists of two parts. The first part, comprising the first two terms, describes natural vibration, that is, the response of the mechanical system to an initial impact in absence of any external excitation afterward. The last term describes the forced response caused by a persistent stimulus, in our case, the motion of the foundation.

The simplest, most efficient, way to study the response of this system to persistent excitation is to use the complex amplitudes as introduced in Section 2.2 of Chapter 2, by assuming the harmonic motion:

$$u_0(t) = u_0 e^{i\omega t}, \quad u(t) = u e^{i\omega t}. \tag{3.7}$$

The complex amplitudes, following the usual convention, are denoted by the same letters as the corresponding time-dependent variables. Upon substituting (3.7) into Eq. (3.5), this equation turns into an algebraic one, and we obtain immediately the relationship between the amplitudes of vibration of the mass and the foundation as functions of frequency:

$$u = \frac{1}{1 - \frac{\omega^2}{\omega_0^2}} u_0. \tag{3.8}$$

The graph of the absolute value of the ratio, $u/u_0$, called the transmissibility function, is shown in Fig. 3.2. It illustrates, in a nutshell, the notion of vibration isolation. As $\omega/\omega_0 \to 0$, $u/u_0 \to 1$, so, at low frequencies vibration is transmitted from the foundation to the mass one-to-one. As $\omega/\omega_0 \to \infty$, $u/u_0 \to 0$, so, at higher frequencies *vibration isolation* takes place: vibration of the mass is less than vibration of the foundation if $\omega > \omega_0\sqrt{2}$. At $\omega = \omega_0$ a *resonance*

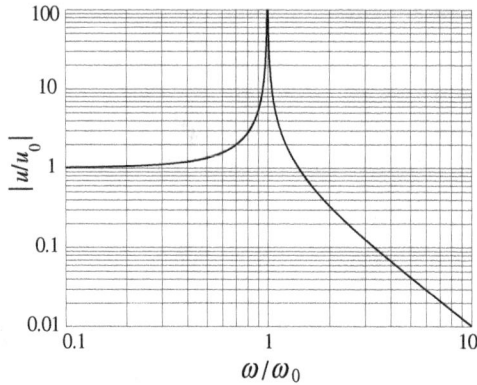

**Figure 3.2.** Transmissibility of the SDOF mechanical oscillator without damping.

occurs: according to Eq. (3.8), $u/u_0 \to \infty$ as $\omega \to \omega_0$. An infinite value of the amplitude is due to the total absence of damping, or dissipation of mechanical energy, in this model. The same assumption explains why the natural vibration, described by the first two terms in Eq. (3.6), continues infinitely without decay.

Section 3.2 introduces dissipation of mechanical energy in the model, which leads to a more realistic behavior.

## 3.2   Simple Linear Oscillator with Damping

This model is a fundamental model of vibration theory. It is amazing how much you can understand in vibration properties of real objects just by understanding this simple model. It differs from our previous model by an additional component representing viscous damping: a linear viscous element depicted by a dashpot in Fig. 3.3. Viscous damping creates resistance proportional in magnitude and opposite in direction to the relative velocity.

Let us consider first the free, or natural, vibration of this system. Assume that the foundation does not move, and no force is acting on the mass. The equations of motion differ from that of the previous section by addition of the viscosity term — a component of force proportional to velocity:

$$m\ddot{u}(t) = F_s, \quad F_s = -ku(t) - c\dot{u}(t).$$

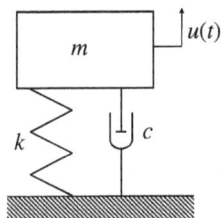

**Figure 3.3.**   SDOF oscillator with linear viscous damping.

Therefore, the equation of motion takes the form

$$m\ddot{u}(t) + c\dot{u}(t) + ku(t) = 0,$$

or, introducing the "undamped" natural frequency (3.3),

$$\ddot{u}(t) + 2\zeta\omega_0\dot{u}(t) + \omega_0^2 u(t) = 0,$$

where

$$\zeta = \frac{c}{2\sqrt{km}}, \tag{3.9}$$

is called *damping ratio*, or *fraction of critical damping*. Introducing the complex amplitude, $u$,

$$u(t) = u e^{i\omega t},$$

gives the following equation for frequency $\omega$:

$$-\omega^2 + 2i\zeta\omega_0\omega + \omega_0^2 = 0.$$

It is called *characteristic equation*. It has two complex solutions:

$$\omega_{1,2} = \pm\omega_0\sqrt{1 - \zeta^2} + i\omega_0\zeta. \tag{3.10}$$

So, the general form of displacement during free vibration as function of time is

$$u(t) = A e^{i\omega_1 t} + B e^{i\omega_2 t}. \tag{3.11}$$

If $\zeta < 1$ (damping less than critical), then Eqs. (3.10) and (3.11) describe oscillatory motion. This is the case normally encountered in vibration control. Assuming $\zeta < 1$, Eq. (3.11) can be re-written in the real form:

$$u(t) = ae^{-\zeta\omega_0 t}\sin\omega_d t + be^{-\zeta\omega_0 t}\cos\omega_d t, \qquad (3.12)$$

where $\omega_d = \omega_0\sqrt{1-\zeta^2}$ is called the *damped* natural frequency. This is the frequency of free vibration of the damped oscillator. Note that $\omega_d < \omega_0$.

**Exercise 3.1.** Find the constants $a$ and $b$ of Eq. (3.12) in terms of initial conditions.

**Solution:** $a = (\dot{u}(0) + \zeta\omega_0 u(0))/\omega_d; b = u(0)$.

Figure 3.4 shows the decaying free vibration of the oscillator due to initial pulse at the equilibrium position (Eq. (3.12) with $b = 0$). The logarithm of the ratio of two successive local peaks of vibration,

$$\Delta = \ln\frac{u_n}{u_{n+1}} = \frac{2\pi\zeta}{\sqrt{1-\zeta^2}} \approx 2\pi\zeta,$$

is called *logarithmic decrement*.

Now let us consider the reaction of this system to the *kinematic excitation*, that is, the vibration of the foundation, in the frequency domain. Following Fig. 3.5, the equation of motion is derived from

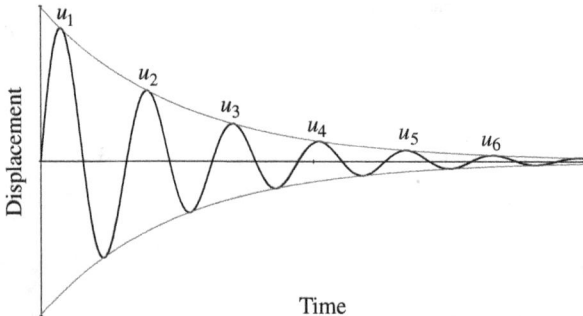

**Figure 3.4.** Decaying natural vibration.

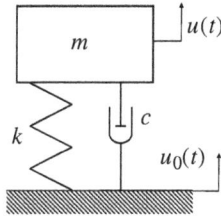

**Figure 3.5.** Damped linear oscillator under kinematic excitation.

the expressions

$$m\ddot{u}(t) = F_s, \quad F_s = -k(u(t) - u_0(t)) - c(\dot{u}(t) - \dot{u}_0(t)),$$

in the following time-domain form:

$$m\ddot{u}(t) + c\dot{u}(t) + ku(t) = c\dot{u}_0(t)) + ku_0(t),$$

and, using (3.7), in complex amplitudes:

$$(-\omega^2 + 2i\zeta\omega_0\omega + \omega_0^2)u = (2i\zeta\omega_0\omega + \omega_0^2)u_0.$$

This gives immediately the complex transmissibility function describing both amplitude and phase relationships between the vibrations of the mass $m$ and the foundation at any frequency:

$$T(\omega) = \frac{u}{u_0} = \frac{\omega_0^2 + 2i\zeta\omega_0\omega}{\omega_0^2 - \omega^2 + 2i\zeta\omega_0\omega} = \frac{1}{1 - \frac{\omega^2}{\omega_0^2\left(1 + 2i\zeta\frac{\omega}{\omega_0}\right)}}. \quad (3.13)$$

The ratio of the real amplitudes of vibration of the mass and the foundation is given by the absolute value of the transmissibility $T(\omega)$:

$$|T(\omega)| = \left|\frac{u}{u_0}\right| = \sqrt{\frac{\omega_0^4 + 4\zeta^2\omega_0^2\omega^2}{(\omega_0^2 - \omega^2)^2 + 4\zeta^2\omega_0^2\omega^2}}. \quad (3.14)$$

This function is plotted in Fig. 3.6 for different values of the damping ratio $\zeta$.

Transmissibility has a maximum at a frequency slightly below $\omega_d$ and, therefore, below $\omega_0$. This frequency is usually referred to as the resonance frequency, although this term is sometimes applied to $\omega_d$. The maximum value of transmissibility is close to $1/2\zeta$ for small values of $\zeta$. The transmissibility graphs of Fig. 3.6 have an invariant

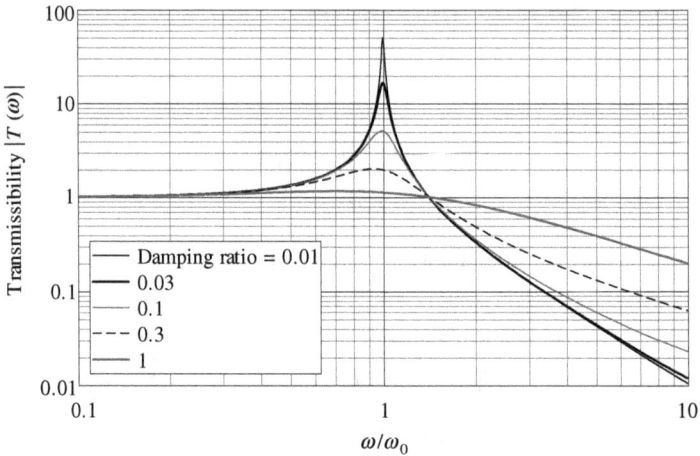

**Figure 3.6.** Transmissibility of the viscously damped oscillator with different values of damping.

point: the curves corresponding to all values of the damping ratio pass through the point $\omega/\omega_0 = \sqrt{2}$, $T = 1$, that marks the transition from the area of resonance amplification at lower frequencies to the area of vibration isolation at higher frequencies. As $\omega \to \infty$, transmissibility decreases asymptotically as $1/\omega$:

$$\omega \to \infty, |T(\omega)| \to 2\zeta\frac{\omega_0}{\omega}. \tag{3.15}$$

This rate of decay is characterized as roll-off by 20 dB per decade or 6 dB per octave.

**Exercise 3.2.** Find the frequency providing maximum value to the transmissibility $|T(\omega)|$ of the viscously damped SDOF oscillator for given $\omega_0$ and $\zeta$.

**Answer:** $\omega_0\sqrt{\dfrac{2}{1+\sqrt{1+8\zeta^2}}}$.

Another loading case for the same model is the force excitation (Fig. 3.7), when the foundation is immovable, and an external force acts directly on the mass. In this case the equation of motion is derived from

$$m\ddot{u}(t) = F + F_s, F_s = -ku(t) - c\dot{u}(t),$$

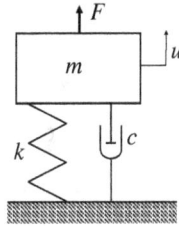

**Figure 3.7.** SDOF oscillator with viscous damping under force excitation.

in the following form:

$$\ddot{u}(t) + 2\zeta\omega_0\dot{u}(t) + \omega_0^2 u(t) = \frac{1}{m}F(t).$$

Introducing the complex amplitudes of force and displacement,

$$F(t) = Fe^{i\omega t}, \quad u(t) = ue^{i\omega t},$$

one obtains the transfer function from force to displacement that characterizes the reaction of the system to the force excitation. This function, along with the transmissibility, plays a key role in vibration control studies. It is called *dynamic compliance* and is expressed, for this model, by the following formula:

$$C(\omega) = \frac{u}{F} = \frac{1}{m(\omega_0^2 - \omega^2 + 2i\zeta\omega_0\omega)} = \frac{1}{k\left(1 - \frac{\omega^2}{\omega_0^2} + 2i\zeta\frac{\omega}{\omega_0}\right)}. \tag{3.16}$$

The absolute value of the dynamic compliance,

$$|C(\omega)| = \frac{|u|}{|F|} = \frac{1}{k\sqrt{\left(1 - \frac{\omega^2}{\omega_0^2}\right)^2 + 4\zeta^2\frac{\omega^2}{\omega_0^2}}}, \tag{3.17}$$

referred to its static value, $1/k$, is plotted in Fig. 3.8 for a range of damping ratios $\zeta$. As seen in Fig. 3.8, the dynamic compliance has a resonance maximum at a certain frequency below $\omega_0$ if $\zeta$ is sufficiently small. There is no invariant point; the compliance rolls off at high frequencies according to the mass law asymptotically as $(-1/m\omega^2)$, or 12 dB per octave, or 40 dB per decade.

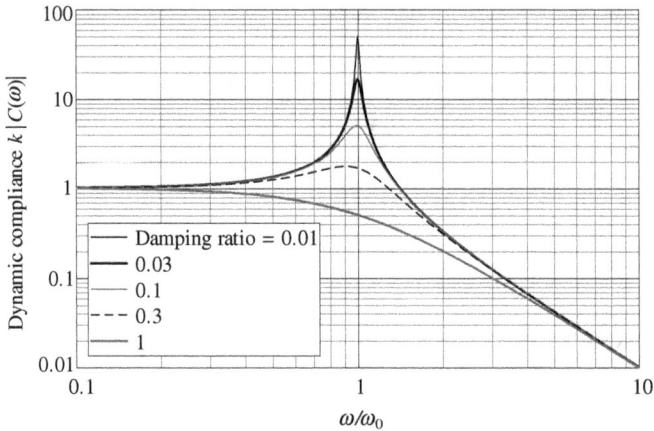

**Figure 3.8.** Dynamic compliance of an SDOF system with viscous damping.

**Exercise 3.3.** Find the frequency providing maximum value to the dynamic compliance $|C(\omega)|$ of the viscously damped SDOF oscillator for given $\omega_0$ and $\zeta$.

**Answer:** $\omega_0\sqrt{1-2\zeta^2}$.

In case of force excitation, transmissibility is defined as the ratio of the force $F_0$ acting on the foundation to the external force $F$, as a function of frequency. This notion is useful for specifying the protection of the surrounding floor from equipment acting as sources of vibration, such as pumps, compressors, electrical motors, etc. Using the following expression for $F_0$,

$$F_0 = k(u(t) - u_0(t)) + c(\dot{u}(t) - \dot{u}_0(t)),$$

it can be shown that the transmissibility, $F_0/F$, is given by the exact same Eq. (3.13):

$$T(\omega) = \frac{F_0}{F} = \frac{\omega_0^2 + 2i\zeta\omega_0\omega}{\omega_0^2 - \omega^2 + 2i\zeta\omega_0\omega} = \frac{1}{1 - \dfrac{\omega^2}{\omega_0^2\left(1 + 2i\zeta\frac{\omega}{\omega_0}\right)}}.$$

So, the force transmissibility is the same as the displacement transmissibility:

$$T(\omega) = \left(\frac{F_0}{F}\right)_{u_0=0} = \left(\frac{u}{u_0}\right)_{F=0}.$$

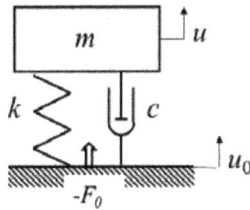

**Figure 3.9.** SDOF oscillator with viscous damping under force excitation from the foundation.

This is a general fact that follows from Rayleigh's reciprocal theorem [1]. It is valid also for MDOF systems.

Yet another loading condition to be considered in connection with this system is the given force acting on the system from the foundation (Fig. 3.9). This case will allow us to derive another frequency characteristic of the damped oscillator (dynamic stiffness at the foundation) and introduce the important concept of antiresonance. The force acting on the system from the foundation, $(-F_0)$, is transmitted directly to the mass; therefore, the motion of the mass and the force $(-F_0)$ are related by Newton's law. In terms of complex amplitudes,

$$(-F_0) = -m\omega^2 u = -m\omega^2 T(\omega)u_0.$$

This defines the dynamic compliance of the system at the foundation:

$$C_0(\omega) = \frac{u_0}{(-F_0)} = -\frac{1}{m\omega^2}\left[1 - \frac{\omega^2}{\omega_0^2\left(1 + 2i\zeta\frac{\omega}{\omega_0}\right)}\right].$$

The absolute value of this function, referred to $1/k$, is plotted in Fig. 3.10. The graphs show that at low levels of damping the dynamic compliance has a sharp minimum at a frequency close to the resonance frequency of the oscillator. The dynamic stiffness at the foundation, $K_0(\omega) = 1/C_0(\omega)$, has a maximum at this frequency. That means that the oscillator has very high resistance to the excitation applied to the damped spring from below. This phenomenon is called antiresonance. It finds wide applications in vibration control, specifically in the design of dynamic vibration absorbers, or tuned mass dampers. It will be discussed further in Chapter 7.

Viscous damping is a very convenient model that adequately describes vibration damping devices based on viscous liquids, such as dashpots. Another model, based on the hysteresis phenomenon

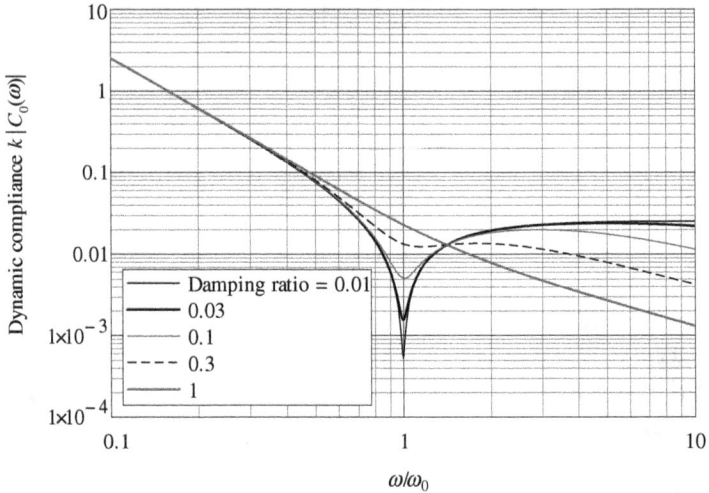

**Figure 3.10.** Dynamic compliance from the side of the foundation. Antiresonance.

in materials, is used for describing materials such as metals, plastics, and elastomers. It leads, with some simplification, to a damping force proportional to the displacement, not the velocity, so that the damping term in the equation of motion in complex amplitudes, say, in case of kinematic excitation, does not depend on frequency:

$$-m\omega^2 u + k(1 + i\eta)(u - u_0) = 0. \tag{3.18}$$

Frequency-independent damping is characterized by the loss factor, $\eta$. Stiffness has an elastic component and a dissipative component, and their ratio is $\eta$. The formula for the transmissibility follows immediately from Eq. (3.18):

$$T(\omega) = \frac{u}{u_0} = \frac{1}{1 - \frac{\omega^2}{\omega_0^2(1+i\eta)}}. \tag{3.19}$$

The absolute value of this function,

$$|T(\omega)| = \left|\frac{u}{u_0}\right| = \sqrt{\frac{\omega_0^4(1 + \eta^2)}{(\omega_0^2 - \omega^2)^2 + (\eta\omega_0^2)^2}},$$

is plotted in Fig. 3.11.

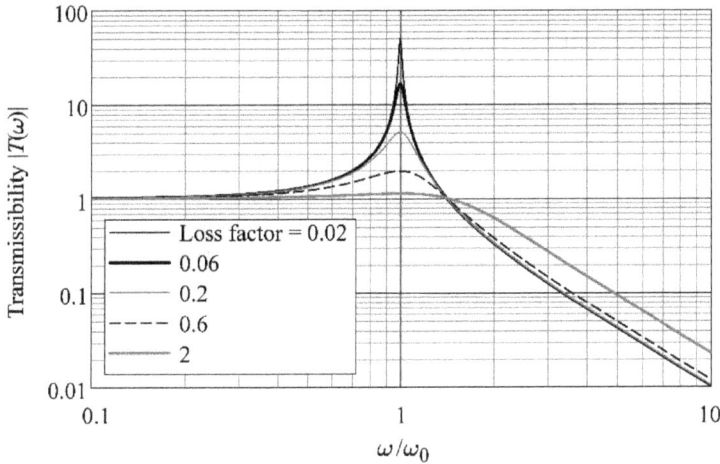

**Figure 3.11.** Transmissibility of an SDOF system with frequency-independent damping.

The transmissibility curves for various values of $\eta$ have maximums at $\omega = \omega_0$, and the invariant point, dividing the isolation zone from the amplification zone, at $\omega = \omega_0\sqrt{2}$. The asymptotic behavior at high frequencies is given by:

$$\omega \to \infty, |T(\omega)| \to \frac{\omega_0^2\sqrt{1+\eta^2}}{\omega^2}.$$

Note that the rate of roll-off at high frequencies is, for this model of damping, 40 dB per decade or 12 dB per octave, compared to 20 dB per decade or 6 dB per octave for the viscous damping model.

**Exercise 3.4.** Find the frequency providing maximum value to the absolute value of the transmissibility $|T(\omega)|$ of the SDOF oscillator with frequency-independent damping for given $\omega_0$ and $\eta$.

**Answer:** $\omega_0$.

**Exercise 3.5.** Derive the expression for the dynamic compliance of the SDOF oscillator with frequency-independent damping for given $\omega_0$ and $\eta$.

**Answer:** $C(\omega) = \frac{u}{F} = \frac{1}{m[\omega_0^2(1+i\eta)-\omega^2]}$.

Caution should be applied at attempts to use this model in the time domain: it can lead to non-causal behavior; corrections should be made to the model at low frequencies [2].

Comparing the levels of damping in viscous and hysteretic models, most attention should be paid to the resonance area where the damping is most important. Equating the dissipative forces at $\omega = \omega_0$, we arrive at the equality

$$\eta = 2\zeta. \tag{3.20}$$

Oftentimes, the notion of loss factor is applied to viscously damped systems, too, in which case it is simply defined by Eq. (3.20).

Viscous and hysteretic models are two simplest models of damping. Adequate modeling of energy dissipation in solids and structures is a non-trivial task that was a subject of a large body of research beyond the scope of this book. Classic monographs [3, 4] give a good introduction to this topic.

For linear systems with arbitrary damping, it is a common practice [4] to represent both stiffness and dissipation properties in the frequency domain by a single complex quantity describing both amplitude and phase of the dynamic reaction: a *complex stiffness*. This model is schematically represented by Fig. 3.12, where a SDOF oscillator includes the mass connected to the foundation by a viscoelastic element combining stiffness and damping. The equation of motion is written in the frequency domain in the following form:

$$-m\omega^2 u + K(\omega)(u - u_0) = 0.$$

Here $K(\omega)$ is the complex stiffness:

$$K(\omega) = \mathrm{Re}[K(\omega)] + i\mathrm{Im}[K(\omega)], \tag{3.21}$$

where the real part is called *storage stiffness*, defining the potential energy stored in the system during a cycle of harmonic vibration, and the imaginary part is called *loss stiffness* and defines the energy

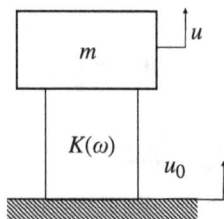

**Figure 3.12.** Oscillator with arbitrary frequency-dependent damping.

dissipated in the system. Their ratio gives the frequency-dependent loss factor:

$$\eta(\omega) = \frac{\text{Im}[K(\omega)]}{\text{Re}[K(\omega)]}.$$

Transmissibility of the system shown in Fig. 3.12 is given by the equation

$$T(\omega) = \frac{K(\omega)}{-m\omega^2 + K(\omega)}, \tag{3.22}$$

which can be considered a generalization of the formulas for vibration transmissibility derived previously in this chapter. Further examples of complex stiffness are considered in Chapter 7 in conjunction with pneumatic and elastomeric vibration isolators.

Transmissibility and dynamic compliance are only two of the family of complex-valued transfer functions commonly used for describing dynamic properties of mechanical systems. Table 3.1 summarizes the relevant terminology assuming a single input (force $F$) and a single output (displacement $u$, velocity $v$, or acceleration $w$). The ANSI/ASA standards [5, 6] detail the definitions and measurement requirements of these transfer functions, both for SDOF and MDOF systems.

As yet another terminological note, the "transmission loss" function, $TL(\omega)$, is sometimes used along with transmissibility, especially in acoustical literature [7]:

$$TL(\omega) = \frac{1}{T(\omega)}.$$

**Table 3.1.** Transfer functions of mechanical systems.

| Transfer Functions | Equations |
|---|---|
| Dynamic compliance | $C = u/F$ |
| Dynamic stiffness | $K = F/u = 1/C$ |
| Mobility or mechanical admittance | $Y = v/F = i\omega C$ |
| Mechanical impedance | $Z = F/v = K/i\omega = 1/Y$ |
| Accelerance | $A = w/F = -\omega^2 C = i\omega Y$ |
| Effective mass or inertance | $M = F/w = -K/\omega^2 = Z/i\omega = 1/A$ |

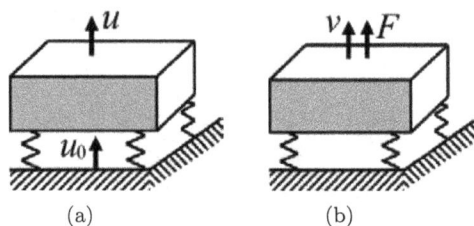

**Figure 3.13.** An isolated platform: (a) kinematic excitation; (b) on-board force excitation.

**Example 3.1. Dynamic reaction of an isolated platform.** Consider a vibration-isolated platform depicted in Fig. 3.13. A SDOF model can be applied to the isolated platform under conditions of symmetry. Suppose the isolators produce the natural frequency of 1 Hz and 15% of critical damping (assuming a linear motion of the isolated platform). So, in frame of the viscous damping model, $\omega_0 = 2\pi \cdot 1$ s$^{-1}$; $\zeta = 0.15$. Assume the mass $m = 1000$ kg. Then the total stiffness of the isolators is $k = m\omega_0^2 = 3.95 \cdot 10^4$ N/m $= 39.5$ N/mm.

(a) **Kinematic excitation from the floor.** Suppose certain rotating equipment, installed on the floor, creates the motion of the floor with the amplitude of 1 $\mu$m and the frequency of 60 Hz. Applying the model of Fig. 3.5, the transmissibility at this frequency ($\omega = 2\pi \cdot 60$ s$^{-1}$) equals, according to Eq. (3.14), $|T(2\pi \cdot 60)| = 5.01 \cdot 10^{-3}$. The asymptotic formula (3.15) gives practically the same result. It is a common practice to assess the severity of vibration by velocity levels. The amplitude of velocity on the floor is $|v_0| = |i\omega u_0| = 2\pi \cdot 60$ s$^{-1} \cdot 1$ $\mu$m $= 377$ $\mu$m/s (note that it is a rather high vibration level, perceptible by humans, see Section 5.2). The amplitude of velocity on the platform is $|v| = |v_0| \cdot |T| = 377$ $\mu$m/s $\cdot 5.01 \cdot 10^{-3} = \mathbf{1.889}$ $\boldsymbol{\mu}$**m/s**. Note that we did not need the mass of the platform or the linear stiffness of isolators for this calculation.

(b) **Force excitation from the on-board equipment.** Suppose a rotational unit with the same frequency of 60 Hz, installed on the isolated mass, has a misbalance of $m_e = 5$ g with the eccentricity of $e = 1$ mm. That would create a harmonic force with the frequency

60 Hz and the amplitude $F_e = m_e \cdot e \cdot (2\pi \cdot f)^2 = 0.005$ kg $\cdot 0.001$ m$\cdot(2\pi \cdot 60$ s$^{-1})^2 = 0.71$ N. Using Eqs. (3.16), (3.17) for the compliance, $C(\omega)$, and the expression for the mobility, $Y(\omega)$, from Table 3.1, one finds, for $\omega = 2\pi \cdot 60$ s$^{-1}$, $|Y(\omega)| = 2.653 \cdot 10^{-6}$. (Note that the asymptotic expressions at $\omega \to \infty$, $C(\omega) \approx -1/m\omega^2$, $Y(\omega) \approx 1/i\omega m$ would lead practically to the same result.) Therefore, the amplitude of velocity of the isolated platform is $|v| = |F_e \cdot Y| = \mathbf{1.885\ \mu m/s}$.

This example is quite realistic, illustrating the fact that on-board excitation of a vibration-isolated platform by a seemingly insignificant source can cause the motion of the platform comparable to the motion transmitted from a very noisy floor.

In conclusion of our brief discussion of SDOF models, let us note an important relationship between the static parameter of the system, namely, its deflection under its own weight, and the dynamic parameter, namely the natural frequency. Since the linear stiffness $k = m\omega_0^2$, the deflection under the static weight load $mg$ is (see Fig. 3.14)

$$\Delta_0 = \frac{g}{\omega_0^2}. \tag{3.23}$$

This equation can be generalized to a wider class of mechanical systems [8]. For plate-like structures,

$$\Delta_0 = C\frac{g}{\omega_0^2},$$

where the constant $C$ can be estimated, approximately, as $C \approx 1.25\ldots1.5$.

**Exercise 3.6.** A load is suspended on a linear spring. How much does the spring stretch under the load if the natural frequency of

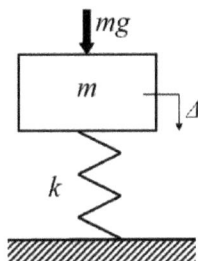

**Figure 3.14.** Deflection of an SDOF system under own weight.

the vertical vibration is 1 Hz (assume the coils of the spring are not pre-stressed)?

**Answer:** 0.25 m.

A SDOF system is an abstract notion: it is hardly encountered in nature and technology in its pure form. Nevertheless, it provides fundamental understanding of the role of main physical properties (inertia, stiffness, and damping) in shaping the oscillatory response of the real-world systems. Besides, the response of complex mechanical systems can be represented by means of modal analysis as a superposition of SDOF modal responses. That will be addressed in Sections 3.4 and 3.5. In some cases, an SDOF model is practically adequate for describing the function of the system; an example is given by the principle of action of vibration sensors (accelerometers and geophones) discussed in Chapter 4.

## 3.3 Oscillators under Random Inputs

In this section, we'll derive useful formulas for root mean square value of vibration displacement of the SDOF oscillator under random excitation (see Section 2.4 of Chapter 2). The oscillator with viscous damping under force excitation is described in the frequency domain by the dynamic compliance (3.16) that we re-write, for the purpose of this section, in terms of frequencies in Hz:

$$H_{uF}(f) = \frac{u}{F} = \frac{1}{m} \frac{1}{(2\pi f_0)^2 \left(1 - \frac{f^2}{f_0^2} + 2i\zeta\frac{f}{f_0}\right)}.$$

According to Eq. (2.10), the power spectral densities of displacement and force are related by

$$W_u(f) = |H_{uF}(f)|^2 W_F(f),$$

where

$$|H_{uF}(f)|^2 = \frac{1}{m^2} \frac{1}{(2\pi f_0)^4 \left[\left(1 - \frac{f^2}{f_0^2}\right)^2 + \left(2\zeta\frac{f}{f_0}\right)^2\right]}.$$

Using Eq. (2.9), one obtains for the RMS displacement

$$u_{\mathrm{RMS}} = \sqrt{\langle u^2 \rangle} = \frac{1}{m(2\pi f_0)^2} \sqrt{\int_0^\infty \frac{W_F(f)\,df}{\left(1 - \frac{f^2}{f_0^2}\right)^2 + \left(2\zeta \frac{f}{f_0}\right)^2}}.$$

Assuming the uniform (white noise) PSD spectrum of force, $W_F(f) = W_F(f_0)$, the integral can be calculated analytically, which gives

$$u_{\mathrm{RMS}} = \frac{\sqrt{W_F(f_0)}}{8\sqrt{\pi^3 f_0^3 \zeta}} \frac{1}{m}. \tag{3.24}$$

Equation (3.24) is valid, approximately, for any input force with a smooth spectrum if the oscillator is low-damped ($\zeta \ll 1$). In this case the vicinity of the resonance frequency is the main contributor to the total RMS value.

In case of kinematic excitation, the complex amplitude of the displacement of the oscillator mass, $u$, is related to that of the foundation, $u_0$, via the transmissibility (3.13):

$$\frac{u}{u_0} = \frac{f_0^2 + 2i\zeta f_0 f}{f_0^2 - f^2 + 2i\zeta f_0 f}.$$

In the same approximation, similar calculations show that the RMS value of dynamic displacement is related to the power spectral density of the displacement of foundation, $W_{u_0}(f)$, by

$$u_{\mathrm{RMS}} = \frac{1}{2} \sqrt{\frac{\pi f_0}{\zeta} W_{u_0}(f_0)}.$$

If the PSD of acceleration on the foundation, $W_{\ddot{u}_0}(f)$ is known, the RMS displacement can be written as

$$u_{\mathrm{RMS}} = \frac{\sqrt{W_{\ddot{u}_0}(f_0)}}{8\sqrt{\pi^3 f_0^3 \zeta}}.$$

If the model of structural (frequency-independent) damping is used, similar calculations under the same assumptions of low damping

$(\eta \ll 1)$ and nearly uniform input PSD lead to analogous formulas, where $\eta/2$ is substituted instead of $\zeta$:

$$u_{\mathrm{RMS}} = \frac{\sqrt{W_f(f_0)}}{\sqrt{32\pi^3 f_0^3 \eta}} \frac{1}{m}, \qquad (3.25)$$

for force excitation and

$$u_{\mathrm{RMS}} = \sqrt{\frac{\pi f_0}{2\eta} W_{u_0}(f)} = \frac{\sqrt{W_{\ddot{u}_0}(f_0)}}{\sqrt{32\pi^3 f_0^3 \eta}}, \qquad (3.26)$$

for kinematic excitation.

The formulas for RMS reaction of a SDOF oscillator to random inputs prove useful in dynamic displacement calculations under random excitation for MDOF and continuous systems via modal analysis that represents vibration of a complex system as a superposition of vibration modes behaving like individual oscillators. It is illustrated in Sections 3.4 and 3.5.

## 3.4　Multi-Degree-of-Freedom Systems

The SDOF model, although of principal importance, may not give enough insight into dynamic behavior of a real object. MDOF systems represent the next level of complexity. We'll start the introduction of MDOF systems with an example of a two-degree-of-freedom (2DOF) system. It is the so-called two-stage system depicted in Fig. 3.15.

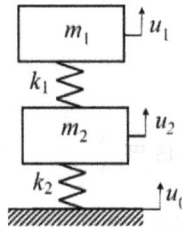

**Figure 3.15.**　2DOF system models two-stage vibration isolation.

The equations of motion are

$$m_1\ddot{u}_1 = -k_1(u_1 - u_2),$$
$$m_2\ddot{u}_2 = k_1(u_1 - u_2) - k_2(u_2 - u_0).$$

In complex amplitudes,

$$(-m_1\omega^2 + k_1)u_1 - k_1 u_2 = 0,$$
$$-k_1 u_1 + (-m_2\omega^2 + k_1 + k_2)u_2 = k_2 u_0. \tag{3.27}$$

Considering the kinematic excitation from the foundation, $u_0$, as an input, and the displacement $u_1$ of the mass $m_1$ as an output, one can resolve (3.27) for the vibration transmissibility:

$$T(\omega) = \frac{u_1}{u_0} = \frac{1}{1 - \left(\frac{m_1}{k_1} + \frac{m_1 + m_2}{k_2}\right)\omega^2 + \frac{m_1 m_2}{k_1 k_2}\omega^4}$$

$$= \frac{1}{\left(1 - \frac{\omega^2}{\omega_1^2}\right)\left(1 - \frac{\omega^2}{\omega_2^2}\right)}.$$

The quadratic function in the denominator has two zeroes, $\omega_1$ and $\omega_2$, which are the natural circular frequencies of the system; therefore, the transmissibility has two resonance peaks. Analysis shows that out of two resonance frequencies of the two-stage system one lies below and one above the "partial" resonance frequencies,

$$\omega_{p1} = \sqrt{\frac{k_1}{m_1}}, \quad \omega_{p2} = \sqrt{\frac{k_2}{m_1 + m_2}}.$$

The damping was not included in this model; to take damping into account, the spring rates $k_1$ and $k_2$ must be considered complex, maybe frequency-dependent according to the applicable damping model. Figure 3.16 compares vibration transmissibilities of two-stage systems with small frequency-independent damping, having the same partial resonance frequencies but different mass ratios $m_2/m_1$. The graphs show that the transmissibility of a two-stage system has faster roll-off at high frequencies compared to the SDOF counterpart. This property is often used in advanced vibration isolation systems: in this case $m_1$ contains the protected optics, and $m_2$ is an intermediate stage introduced for improving high-frequency isolation. Sometimes even more stages are employed (an example is considered in

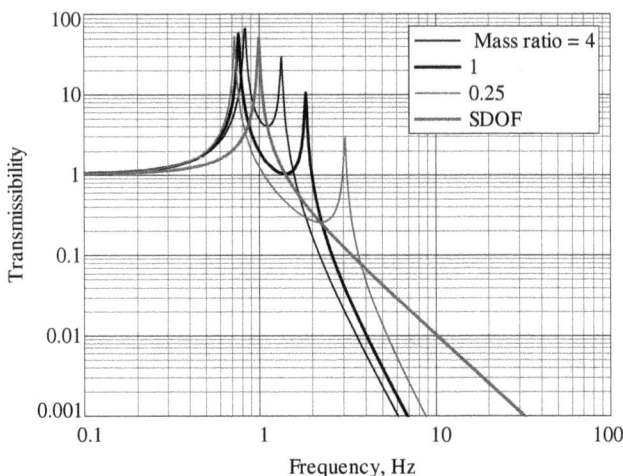

**Figure 3.16.** Transmissibilities of optimal two-stage isolation systems with different mass ratios $m_2/m_1$; $\eta = 0.02$; $\omega_{p1} = \omega_{p2} = 2\pi \cdot 1$ Hz.

Section 9.3 of Chapter 9). The downside is the expanded resonance zone with more than one resonance peak.

Two-stage vibration isolation can be optimized in the sense of minimizing this resonance domain, that is, minimizing the ratio $\omega_2/\omega_1$. The minimum possible value of this ratio is

$$\left(\frac{\omega_2}{\omega_1}\right)_{\min} = \sqrt{\frac{\sqrt{m_2 + m_1} + \sqrt{m_1}}{\sqrt{m_2 + m_1} - \sqrt{m_1}}}.$$

The minimum is achieved when the stiffness is distributed between the stages according to the following condition:

$$\frac{k_1}{k_2} = \frac{m_1}{m_1 + m_2},$$

that is, $\omega_{p1} = \omega_{p2}$. This means that in an optimal two-stage isolation system of Fig. 3.15 the partial frequency of the upper stage equals the natural frequency of the system with the upper stage docked. The systems referenced in Fig. 3.16 have this property. Details of optimization can be found in [9, 10].

The same model can be used to estimate the influence of the internal resonances of vibration isolation systems. If the mass $m_2$ represents internal inertial properties of the isolator, it can manifest itself as an additional resonance peak; since the mass of the isolator is usually much smaller than the mass of the isolated object, this resonance will show somewhere down the rolled-off area of the transmissibility curve.

Another 2DOF model of principal importance to vibration control is the model of dynamic vibration absorber considered in detail in Section 7.2 of Chapter 7.

Moving to more complex mechanical systems, consider a system of $n$ material points elastically connected with each other and with the foundation (Fig. 3.17). This model can be envisioned as a universal model in theoretical mechanics. For sake of simplicity and physical transparency, assume that each material point can move in only one direction coinciding with the direction of the external excitation [9, 11].

The equations of motion of such an MDOF system are most conveniently formulated in a matrix form. Generalized coordinates (degrees of freedom) of the system form an $n$-dimensional vector $\mathbf{u}$ comprised of displacement of individual material points,

$$\mathbf{u} = \left\{ \begin{array}{c} u_1 \\ \vdots \\ u_n \end{array} \right\}.$$

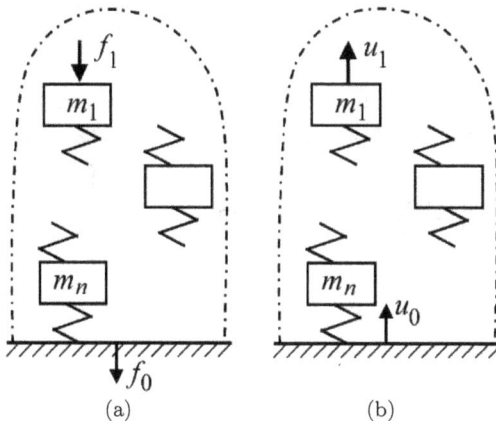

**Figure 3.17.** MDOF system consisting of n material points: (a) force excitation; (b) kinematic excitation.

Kinetic energy of the system, $T_K$, and potential energy, $U$, are quadratic forms of displacements defined by matrices $\mathbf{M}$ and $\mathbf{C}$:

$$T_K = \frac{1}{2}\dot{\mathbf{u}}^T\mathbf{M}\dot{\mathbf{u}}, \quad U = \frac{1}{2}\mathbf{u}^T\mathbf{C}\mathbf{u},$$

$$\mathbf{C} = \begin{bmatrix} c_{11} & \cdots & c_{1n} \\ & \ddots & \\ c_{1n} & \cdots & c_{nn} \end{bmatrix}, \mathbf{M} = \begin{bmatrix} m_1 & & 0 \\ & \ddots & \\ 0 & \cdots & m_n \end{bmatrix}.$$

Equation of motion under the force excitation, Fig. 3.17(a), is written in the matrix form as follows:

$$\mathbf{M}\ddot{\mathbf{u}}(t) + \mathbf{C}\mathbf{u}(t) = \mathbf{f}(t).$$

Here $\mathbf{f}(t)$ is the vector of the external forces. In complex amplitudes, using the same notation for complex amplitudes and time-dependent quantities, the equation of motion takes the matrix algebraic form:

$$(\mathbf{C} - \omega^2\mathbf{M})\mathbf{u} = \mathbf{f}, \tag{3.28}$$

and the solution is

$$\mathbf{u} = (\mathbf{C} - \omega^2\mathbf{M})^{-1}\mathbf{f}.$$

To derive the expression for the transmissibility of this MDOF system, assume that the external force is acting on $m_1$ as shown in Fig. 3.17(a):

$$\mathbf{f} = \begin{Bmatrix} f_1 \\ 0 \\ \vdots \\ 0 \end{Bmatrix} = f_1\mathbf{e}_1, \quad \mathbf{e}_1 = \begin{Bmatrix} 1 \\ 0 \\ \vdots \\ 0 \end{Bmatrix}.$$

The output force on the foundation, $f_0$, can be calculated by adding up all elastic forces so that the internal forces cancel each other:

$$f_0 = \sum c_{ij}u_j = \mathbf{e}^T\mathbf{C}\mathbf{u}, \quad \mathbf{e} = \begin{Bmatrix} 1 \\ \vdots \\ 1 \end{Bmatrix}.$$

Therefore, vibration transmissibility is expressed by the following equation:

$$T(\omega) = \frac{f_0}{f_1} = \mathbf{e}^T\mathbf{C}(\mathbf{C} - \omega^2\mathbf{M})^{-1}\mathbf{e}_1. \tag{3.29}$$

In case of kinematic excitation (Fig. 3.17(b)), the equations of motion are written in the coordinate system moving with the foundation, using D'Alembert's principle that requires adding inertia forces to the right-hand side of the equation:

$$\mathbf{M}[\ddot{\mathbf{u}}(t) - \ddot{\mathbf{u}}_0(t)] - \mathbf{C}[\mathbf{u}(t) - \mathbf{u}_0(t)] = -\mathbf{M}\ddot{\mathbf{u}}_0(t),$$

where $\mathbf{u}_0 = u_0\mathbf{e}$. In complex amplitudes, the equation of motion takes the following algebraic form:

$$(\mathbf{C} - \omega^2\mathbf{M})\mathbf{u} = u_0\mathbf{C}\mathbf{e},$$

so that

$$\mathbf{u} = (\mathbf{C} - \omega^2\mathbf{M})^{-1}\mathbf{C}\mathbf{e}u_0,$$

and the vibration transmissibility,

$$T(\omega) = \frac{u_1}{u_0} = \mathbf{e}_1^T(\mathbf{C} - \omega^2\mathbf{M})^{-1}\mathbf{C}\mathbf{e}, \tag{3.30}$$

is equal to (3.29) for symmetric matrices $\mathbf{C}$ and $\mathbf{M}$, in accordance with Rayleigh's reciprocal theorem [1].

The easy way to include damping is to consider the stiffness matrix as consisting of real "storage" matrix and imaginary "loss" matrix:

$$\mathbf{C} = \mathbf{C}_s + i\mathbf{C}_l.$$

In the simplest case of uniform frequency-independent damping, the two matrices are proportional:

$$\mathbf{C} = \mathbf{C}_s(1 + i\eta). \tag{3.31}$$

We assumed that the model is "unidirectional" for sake of simplicity. If the input and output vectors are three-dimensional, the vibration transmissibility can be derived in the same manner as a $(3 \times 3)$ frequency-dependent matrix [9].

Modal analysis is a convenient way to study the vibrational behavior of MDOF systems. It is based on the representation of the dynamic response $\mathbf{u}$ of a system with $n$ degrees of freedom as a superposition of $n$ vibration modes $\varphi$, each corresponding to natural vibration with one of the natural frequencies, $\omega_k$:

$$\mathbf{C}\varphi_k - \omega_k^2 \mathbf{M}\varphi_k = 0, \quad k = 1, 2 \ldots, n.$$

Natural frequencies satisfy the characteristic equation

$$|\mathbf{C} - \omega^2 \mathbf{M}| = 0. \tag{3.32}$$

Matrix $\boldsymbol{\Phi}$, composed of columns $\varphi_k$, reduces both mass matrix and stiffness matrix to diagonal form:

$$\boldsymbol{\Phi}^{\mathrm{T}}\mathbf{M}\boldsymbol{\Phi} = \mathbf{E}, \quad \boldsymbol{\Phi}^{\mathrm{T}}\mathbf{C}\boldsymbol{\Phi} = \boldsymbol{\Lambda},$$

where $\mathbf{E}$ is a unity matrix, $\boldsymbol{\Lambda} = \mathrm{diag}\{\omega_1^2, \ldots, \omega_n^2\}$. The natural frequency squared can be expressed as the ratio of potential energy and kinetic energy quadratic forms associated with the mode:

$$\omega_k^2 = \frac{\varphi_k^{\mathrm{T}} \mathbf{C} \varphi_k}{\varphi_k^{\mathrm{T}} \mathbf{M} \varphi_k}. \tag{3.33}$$

Equation (3.33) and the corresponding minimum principle form the basis for the theorem stating that the natural frequencies increase when the stiffness of the system increases and decrease when the mass increases. Also, the natural frequencies increase if additional constraints are imposed on the system [1].

The vibration modes are orthogonal with respect to the mass matrix; they are often normalized with respect to this matrix (physically it means normalization with respect to the total kinetic energy):

$$\varphi_k^{\mathrm{T}} \mathbf{M} \varphi_j = 0, \quad k \neq j; \quad \varphi_k^{\mathrm{T}} \mathbf{M} \varphi_k = 1. \tag{3.34}$$

Then the complex amplitudes of dynamic displacements of the undamped system under force excitation are given by

$$\mathbf{u} = \sum_{k=1}^{n} \frac{\mathbf{f}^{\mathrm{T}} \varphi_k}{\omega_k^2 - \omega^2} \varphi_k \tag{3.35}$$

for normalized modes (3.34). The dynamic displacement of a damped system can be approximated by the modal expansion using the complex natural frequencies derived from complex stiffnesses; more details on this can be found elsewhere [9, 11]. For the complex stiffness matrix of the special form (3.31), the complex amplitudes of the displacement are

$$\mathbf{u} = \sum_{k=1}^{n} \frac{\mathbf{f}^{\mathrm{T}} \boldsymbol{\varphi}_k}{\omega_k^2 (1 + i\eta) - \omega^2} \boldsymbol{\varphi}_k, \tag{3.36}$$

where $\omega_k$ are undamped natural frequencies. Expansions (3.35) and (3.36) assume that the modes are normalized by the second equation (3.34). For small non-uniform damping, approximate representation of displacement can be obtained by assuming individual modal loss factors, $\eta_k$, in each member of the sum (3.36).

**Exercise 3.7.** If the MDOF system vibrates at its natural frequency $\omega_k$, and the vector of amplitudes follows exactly the natural mode $\boldsymbol{\varphi}_k$, normalized by (3.33), what is the numerical value of the kinetic energy?

**Answer:** $\omega_k^2/2$.

Note the expression for the local dynamic compliance:

$$C_j(\omega) = \frac{u_j}{F_j} = \sum_{k=1}^{n} \frac{1}{m_{jk}} \frac{1}{\omega_k^2 (1 + i\eta_k) - \omega^2}, \tag{3.37}$$

where each term under the sum has the same form as dynamic compliance of a single-degree-of-freedom oscillator (see Exercise 3.5), and $m_{jk} = 1/\varphi_{jk}^2$ can be interpreted as an equivalent modal mass. Recalling Section 3.3 (specifically, Eq. (3.25)), one obtains the following estimate of the RMS displacement of the $j$th mass modal response caused by the random input in the vicinity of the $k$th natural frequency, as defined by the given power spectral density of the input force $W_F(f)$:

$$[(u_j)_{\mathrm{RMS}}]_k = \frac{\sqrt{W_F(f_k)}}{\sqrt{32\pi^3 f_k^3 \eta_k}} \frac{1}{m_{jk}}, \tag{3.38}$$

where $f_k = \omega_k/2\pi$ are natural frequencies in Hz. Mean square of the total response is closely approximated, at least for small damping, by the sum of modal contributions [12]:

$$\langle u_j^2 \rangle = \sum_{k-1}^{n} \frac{W_F(f_k)}{32\pi^3 f_k^3 \eta_k} \frac{1}{m_{jk}^2}.$$

One case of MDOF system deserves special attention because of its practical importance. This is the case of a rigid body elastically supported by springs (vibration-isolated) as shown in Fig. 3.18. It has six degrees of freedom. There is an arbitrary number of $n$ springs. Their attachment points to the rigid body are defined by coordinates $(x_i, y_i, z_i)$ in the orthogonal coordinate system $Oxyz$, $i = 1, 2, \ldots, n$. Assume further that all springs (vibration isolators) have distinct vertical and horizontal stiffnesses. (More general case of inclined springs can be found in the handbook [13].)

The vertical $w_i$ and horizontal $u_i$, $v_i$ displacements at the support points are defined by the displacements of the origin $u_0$, $v_0$, $w_0$, and rotations around the coordinate axes $\theta = (\theta_x, \theta_y, \theta_z)$:

$$\begin{aligned}
u_i &= u_0 + z_i\theta_y - y_i\theta_z, \\
v_i &= v_0 - z_i\theta_x + x_i\theta_z, \\
w_i &= w_0 + y_i\theta_x - x_i\theta_y.
\end{aligned} \tag{3.39}$$

The potential energy is

$$U = \frac{1}{2}\sum_i k_i^{(v)} w_i^2 + \frac{1}{2}\sum_i k_i^{(h)}(u_i^2 + v_i^2).$$

The kinetic energy is written down in terms of linear and angular velocities as the sum of the energy of the linear motion of the center

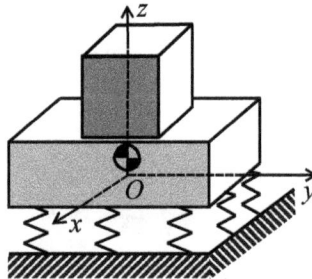

**Figure 3.18.** Elastically supported rigid body.

of gravity and the energy of the rotational motion around the center of gravity:

$$T_k = \frac{1}{2}m(\dot{u}_c^2 + \dot{v}_c^2 + \dot{w}_c^2) + \frac{1}{2}\dot{\boldsymbol{\theta}}\mathbf{J}\dot{\boldsymbol{\theta}}^{\mathrm{T}},$$

where the displacements of the center of gravity (CG), $u_c$, $v_c$, $w_c$, are expressed through the coordinates of the CG ($x_c$, $y_c$, $z_c$) by equations analogous to (3.39):

$$u_c = u_0 + z_c\theta_y - y_c\theta_z,$$
$$v_c = v_0 - z_c\theta_x + x_c\theta_z,$$
$$w_c = w_0 + y_c\theta_x - x_c\theta_y,$$

$m$ is the total supported mass, and $\mathbf{J}$ is the matrix of the moments of inertia at the CG:

$$\mathbf{J} = \begin{pmatrix} J_x & -J_{xy} & -J_{xz} \\ -J_{xy} & J_y & -J_{yz} \\ -J_{xz} & -J_{yz} & J_z \end{pmatrix}.$$

(See textbooks [14, 15] for more information on the inertia properties of a rigid body.) Kinetic and potential energies can be, again, represented as quadratic forms of the generalized coordinates $\mathbf{u} = (u_0, v_0, w_0, \theta_x, \theta_y, \theta_z)$, which leads to the equations of motion in the form (3.28), $\mathbf{f} = (F_x, F_y, F_z, M_x, M_y, M_z)$ being the vector of generalized forces (total forces and moments) acting on the rigid body, with the stiffness and inertia matrices (3.40).

The natural frequencies are calculated from the characteristic equation (3.32). The transmissibilities are given by the equation analogous to (3.29), (3.30). The full set of transmissibilities forms a $6 \times 6$ matrix $\mathbf{T}(\omega)$ according to Eq. (3.41).

$$\mathbf{C} =$$

$$\begin{pmatrix}
\sum_i k_i^{(h)} & 0 & 0 & 0 & \sum_i k_i^{(h)} z_i & -\sum_i k_i^{(h)} y_i \\
0 & \sum_i k_i^{(h)} & 0 & -\sum_i k_i^{(h)} z_i & 0 & \sum_i k_i^{(h)} x_i \\
0 & 0 & \sum_i k_i^{(v)} & \sum_i k_i^{(v)} y_i & -\sum_i k_i^{(v)} x_i & 0 \\
0 & -\sum_i k_i^{(h)} z_i & \sum_i k_i^{(v)} y_i & \sum_i (k_i^{(v)} y_i^2 + k_i^{(h)} z_i^2) & \sum_i k_i^{(v)} x_i y_i & -\sum_i k_i^{(h)} x_i z_i \\
\sum_i k_i^{(h)} z_i & 0 & -\sum_i k_i^{(v)} x_i & \sum_i k_i^{(v)} x_i y_i & \sum_i (k_i^{(v)} x_i^2 + k_i^{(h)} z_i^2) & -\sum_i k_i^{(h)} y_i z_i \\
-\sum_i k_i^{(h)} y_i & \sum_i k_i^{(h)} x_i & 0 & -\sum_i k_i^{(h)} x_i z_i & -\sum_i k_i^{(h)} y_i z_i & \sum_i (k_i^{(h)} x_i^2 + k_i^{(h)} y_i^2)
\end{pmatrix}$$

$$\mathbf{M} =$$

$$\begin{pmatrix} m & 0 & 0 & 0 & mz_c & -my_c \\ 0 & m & 0 & -mz_c & 0 & mx_c \\ 0 & 0 & m & my_c & -mx_c & 0 \\ 0 & -mz_c & my_c & J_x + m(z_c^2 + y_c^2) & -J_{xy} - mx_c y_c & -J_{xz} - mx_c z_c \\ mz_c & 0 & -mx_c & -J_{xy} - mx_c y_c & J_y + m(x_c^2 + z_c^2) & -J_{yz} - my_c z_c \\ -my_c & mx_c & 0 & -J_{xz} - mx_c z_c & -J_{yz} - my_c z_c & J_z + m(y_c^2 + x_c^2) \end{pmatrix}. \quad (3.40)$$

$$\mathbf{T}(\omega) = (\mathbf{C} - \omega^2 \mathbf{M})^{-1} \mathbf{C}. \quad (3.41)$$

So, if the foundation (assumed to be rigid) vibrates harmonically with complex amplitudes $\mathbf{u}_b = (u_b, v_b, w_b, \varphi_x^b, \varphi_y^b, \varphi_z^b)$, the motion at the origin $O$ is characterized by generalized displacements

$$\mathbf{u} = [\mathbf{T}(\omega)]\mathbf{u}_b.$$

(Note that the effects of centripetal and Coriolis accelerations are neglected in the linear approximation.) Accordingly, if the rigid body is subject to external forces and moments with complex amplitudes $\mathbf{f} = (F_x, F_y, F_z, M_x, M_y, M_z)$, and the foundation is fixed, the resulting generalized force, $\mathbf{f}_b = (F_x^b, F_y^b, F_z^b, M_x^b, M_y^b, M_z^b)$, on the foundation, is

$$\mathbf{f}_b = [\mathbf{T}(\omega)]^{\mathrm{T}} \mathbf{f} = \mathbf{C}(\mathbf{C} - \omega^2 \mathbf{M})^{-1} \mathbf{f}.$$

The principle of reciprocity is still valid: for example, the $x$ — motion at the origin $O$ per unit of kinematic excitation on the foundation in the direction $z$, is the same as the reaction force on the foundation in the direction $z$ caused by the unit of force excitation at the origin $O$ in the direction $x$, both being equal to $T_{13}$.

To find the transmissibility $\mathbf{T}_a$ to/from an arbitrary point $(x_a, y_a, z_a)$, the transformation matrix is used:

$$\mathbf{T}_a(\omega) = \mathbf{R}_a \mathbf{T}(\omega); \mathbf{R}_a = \begin{pmatrix} 1 & 0 & 0 & 0 & z_a & -y_a \\ 0 & 1 & 0 & -z_a & 0 & x_a \\ 0 & 0 & 1 & y_a & -x_a & 0 \\ 0 & 0 & 0 & 1 & 0 & 0 \\ 0 & 0 & 0 & 0 & 1 & 0 \\ 0 & 0 & 0 & 0 & 0 & 1 \end{pmatrix}. \quad (3.42)$$

Damping can be, in the first approximation, included by introducing complex-valued stiffnesses. Note that this model takes into account only the potential energy of the deformation of the springs. The dynamic behavior is more complicated for supported structures with high positions of the center of gravity, when the changes in the potential energy of gravity must be considered that contribute to (or, rather, detract from) effective stiffness. We'll address this phenomenon in Chapter 6.

**Exercise 3.8. Transmissibilities of an elastically supported honeycomb sandwich platform.** A rectangular platform, $L = 2.5$ m long, $W = 1.25$ m wide and $H = 0.3$ m thick, consists of steel skins 5 mm thick each and steel honeycomb core with a 2.5% volume ratio. The platform is supported by four vibration isolators in a symmetric pattern with spans $L_1 = 2.2$ m in the length direction and $W_1 = 1$ m in the width direction (see Fig. 3.19). Isolators can be modeled by linear springs possessing distinct vertical and horizontal stiffnesses with nominal frequencies $f_v = 1.2$ Hz and $f_h = 1.5$ Hz in vertical and horizontal directions respectively. This structure is typical for state-of-the-art vibration isolated platforms such as optical tables (see a more detailed description of optical tables in Chapters 6 and 7). Considering the platform as a rigid body, find the natural frequencies and plot the vibration transmissibilities in vertical and horizontal directions.

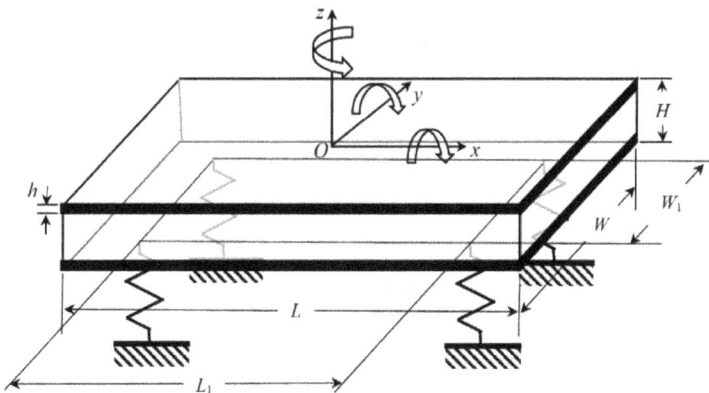

**Figure 3.19.** Geometry of elastic supports and the coordinate system of Exercise 3.8 (sketch not to scale).

**Solution:** Assuming the mass density of steel $\rho_s = 7.8 \cdot 10^3$ kg/m$^3$, the total mass $m = m_{skin} + m_{core} = 2\rho_s LWh + 0.025\rho_s LW(H - 2h) = 420.5$ kg. Individual stiffness rates of isolators $k^{(v)} = \frac{1}{4}m(2\pi f_v)^2 = 5976$ N/m, $k^{(h)} = \frac{1}{4}m(2\pi f_h)^2 = 9337$ N/m.

Moments of inertia, calculated by standard formulas [13], equal (in kg·m$^2$):

$$J_x = m_{skin}\left[\left(\frac{H}{2} - h\right)\frac{H}{2} + \frac{h^2}{3} + \frac{W^2}{12}\right]$$
$$+ m_{core}\left[\frac{W^2}{12} + \frac{1}{3}\left(\frac{H}{2} - h\right)^2\right] = 61.29,$$

$$J_y = m_{skin}\left[\left(\frac{H}{2} - h\right)\frac{H}{2} + \frac{h^2}{3} + \frac{L^2}{12}\right]$$
$$+ m_{core}\left[\frac{L^2}{12} + \frac{1}{3}\left(\frac{H}{2} - h\right)^2\right] = 225.54,$$

$$J_z = m\frac{(W^2 + L^2)}{12} = 273.74, \quad J_{xy} = J_{xz} = J_{yz} = 0.$$

The origin of coordinates is placed at the center of the top surface of the platform. The coordinates of the elastic supports are $x_1 = x_2 = L_1/2 = 1.1$ m, $x_3 = x_4 = -L_1/2 = -1.1$ m; $y_1 = y_4 = -W_1/2 = -0.5$ m, $y_2 = y_3 = W_1/2 = 0.5$ m; $z_1 = z_2 = z_3 = z_4 = -H = -0.3$ m; the coordinates of the CG are $x_c = y_c = 0$, $z_c = -H/2 = -0.15$ m.

This leads to the following numerical values for matrices of stiffness and inertia (3.40):

$$C = \begin{pmatrix}
3.375 & 0 & 0 & 0 & -1.12 & 0 \\
0 & 3.375 & 0 & 1.12 & 0 & 0 \\
0 & 0 & 2.39 & 0 & 0 & 0 \\
0 & 1.12 & 0 & 9.337 & 0 & 0 \\
-1.12 & 0 & 0 & 0 & 3.228 & 0 \\
0 & 0 & 0 & 0 & 0 & 5.453
\end{pmatrix} \cdot 10^4,$$

**Table 3.2.**   Natural frequencies and modes of the elastically supported rigid body of Exercise 3.8.

| Mode number | 1 | 2 | 3 | 4 | 5 | 6 |
|---|---|---|---|---|---|---|
| Frequency, Hz | 1.2 | 1.267 | 1.443 | 1.861 | 1.874 | 2.246 |
| Degrees-of-freedom involved | $z$ | $y$ and $\theta_x$ | $x$ and $\theta_y$ | $y$ and $\theta_x$ | $x$ and $\theta_y$ | $\theta_z$ |

$$
\mathbf{M} = \begin{pmatrix}
420.47 & 0 & 0 & 0 & -63.07 & \\
0 & 420.47 & 0 & 63.07 & 0 & 0 \\
0 & 0 & 420.47 & 0 & 0 & 0 \\
0 & 63.07 & 0 & 70.75 & 0 & 0 \\
-63.07 & 0 & 0 & 0 & 234.0 & 0 \\
0 & 0 & 0 & 0 & 0 & 273.74
\end{pmatrix}.
$$

The natural frequencies are found by numerical solution of the characteristic equation or by application of any available standard routine for the eigenproblem solution. The resulting natural frequencies are listed in Table 3.2:

The results show that unimodal natural vibrations exist representing linear vertical motion and rotation around the vertical axis. This is due to the symmetry of the system. However, since the center of gravity lies off the support plane, the system has mixed modes combining horizontal motion with rotation around the perpendicular horizontal axis. If the foundation experiences motion in the horizontal direction, the platform reacts with both horizontal motion and rocking motion. It is illustrated by the transmissibility curves calculated by Eq. (3.41) and plotted in Figs. 3.20 and 3.21. Vertical transmissibility is perfectly unimodal; horizontal transmissibilities in $x$ and, especially, in $y$-direction, show the second peak reflecting the rocking motion. Rocking motion excited by horizontal movement of the floor leads to vertical motion on the surface of the table away from its center. A constant loss factor of $\eta = 0.1$ was assumed in these calculations.

Note that in the general case of non-symmetric distribution of payload on the platform, the transmissibility can be a full $6 \times 6$ matrix, so that excitation in any direction, linear or rotational, can lead to reactions in all six rigid-body degrees-of-freedom. At higher

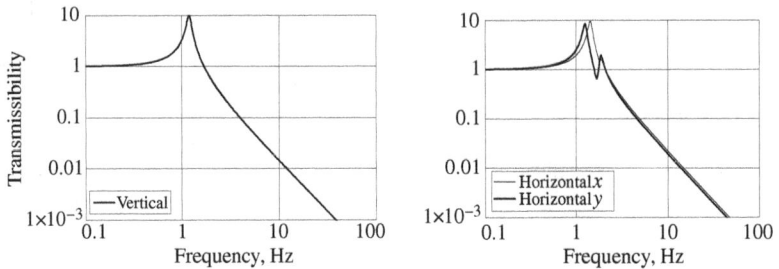

**Figure 3.20.** Vibration transmissibilities from the floor to the origin (top center of the platform) in three coordinate directions.

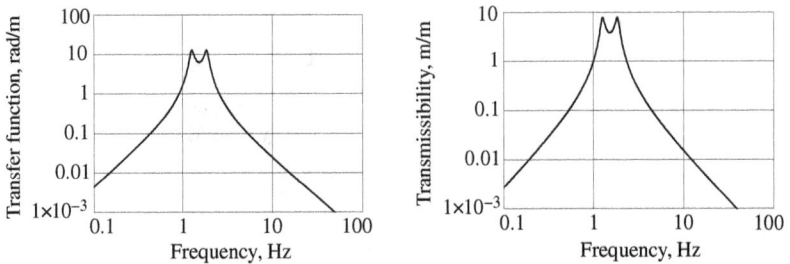

**Figure 3.21.** Rocking of the platform due to horizontal ($y$) motion of the floor: (a) rotation around the $x$-axis, (b) vertical displacement (per unit displacement in $y$) at a corner of the platform.

frequencies, an isolated platform can behave as a non-rigid body, and more complicated (continuous) models must be employed for describing its dynamic properties.

## 3.5　Continuous Systems

Continuous models are often used for more detailed analysis of vibration control systems and optomechanics. They include various models of beams, plates, shells, and solid bodies described by ordinary or partial differential equations.

An elastic beam is a simplest continuous model that can be used for analyzing dynamical behavior of optomechanical structures such as optical posts, post holders, bridges, cages, and frames. The model of elastic beam describes the structure with one dimension much larger than the other two, such as narrow plates, rods, beams, and columns. Flexural vibration of a thin beam with a symmetric cross-section in its plane of symmetry is governed by the differential equation (see textbooks [14, 15] for details):

$$\rho A \ddot{u}(x,t) = -\frac{\partial^2}{\partial x^2}\left(EI\frac{\partial^2 u(x,t)}{\partial x^2}\right) + p(x,t),$$

where $u$ is the lateral displacement, $E$ is Young's modulus, $I$ is the moment of inertia of the cross-section, $\rho$ is the mass density, $A$ is the area of the cross-section, $p$ is the lateral force per unit length; $x$ is the coordinate along the beam's central axis (see Fig. 3.22). This equation is based on Bernoulli–Euler theory, which assumes that the cross-sections of the beam stay flat and orthogonal to the centerline. More detailed models exist, useful mostly at higher frequencies and for shorter beams [15].

Constitutive equation for a Bernoulli–Euler beam relates the moment $M$ acting on the beam's cross-section to the curvature of the beam axis, equal, in the linear approximation, to the second derivative of the displacement:

$$M = EI\frac{\partial^2 u}{\partial x^2}. \tag{3.43}$$

The shear force in the cross-section is

$$Q = \frac{\partial M}{\partial x}. \tag{3.44}$$

The sign convention is illustrated in Fig. 3.22.

To study the vibrational properties of the beam, we introduce, as in Section 3.1, a complex amplitude that is now a function of

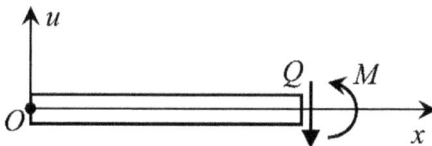

**Figure 3.22.** Beam model. Coordinate system and sign convention.

coordinate $x$, and denote it by the same letter for sake of simplicity:

$$u(x,t) = u(x)e^{i\omega t}.$$

In absence of the external forces ($p = 0$) this leads to the equation for the natural modes of the beam,

$$\frac{d^2}{dx^2}\left(EI\frac{d^2u(x)}{dx^2}\right) = \rho A\omega^2 u(x), \tag{3.45}$$

that must be complemented by four boundary conditions describing supports or constraints imposed at both ends. Non-trivial solutions of this equation and boundary conditions define natural frequencies and natural modes. The solution for a uniform beam with symmetric constant cross-section ($EI = $ const and $\rho A = $ const) gives the following formula defining natural frequencies of bending vibration:

$$\omega_n = \beta^2 \frac{1}{L^2}\sqrt{\frac{EI}{\rho A}},$$

where $L$ is the length of the beam and $\beta$ is the constant depending on the boundary conditions as shown in Fig. 3.23. Detailed analysis of these cases can be found in numerous textbooks and handbooks [14, 15, 17]. Applications of the beam model are illustrated below in Examples 3.2–3.5.

**Example 3.2. Natural frequencies of an optical post.** A steel optical post with diameter $d = 25$ mm and height $L = 150$ mm carries an optomechanical payload, characterized by the mass $m = 200$ g and moment of inertia $J = 200$ kg$\cdot$mm$^2$ at the attachment point (Fig. 3.24). Find the natural frequencies of the system assuming rigid attachment of the post to the base and the payload.

Consider the optical post as a cantilever uniform Bernoulli–Euler beam. Further, assume that the motion is restricted to the plane of the drawing. For a uniform beam, Eq. (3.45) can be re-written as

$$\frac{d^4u(x)}{dx^4} - \kappa^4 u(x) = 0, \tag{3.46}$$

where

$$\kappa^4 = \frac{\rho A}{EI}\omega^2. \tag{3.47}$$

The general solution of this equation is a linear combination of functions $\sin\kappa x$, $\cos\kappa x$, $\sinh\kappa x$, $\cosh\kappa x$. The boundary conditions at

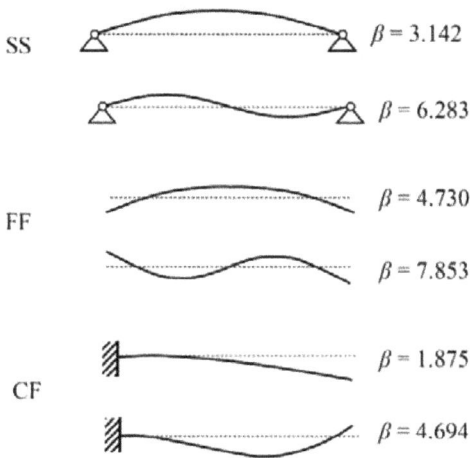

**Figure 3.23.** Natural modes and frequency constants of a uniform elastic beam under different boundary conditions: simply supported on both sides, free-free and clamped-free.

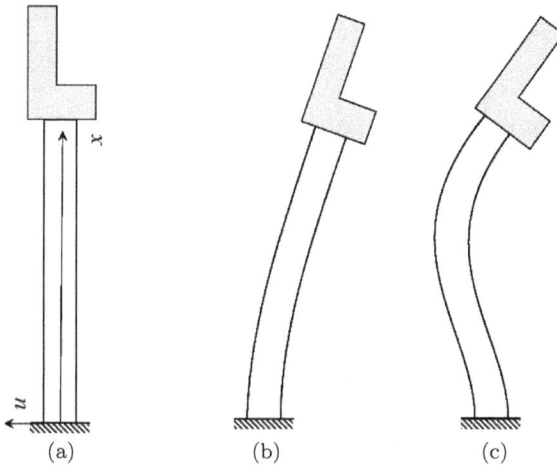

**Figure 3.24.** Example 3.2: Coordinate system (a), first mode (b), second mode (c). Drawing not to scale.

$x = 0$ are $u(0) = 0$, $u'(0) = 0$, which allows to put this general solution in the convenient form

$$u(x) = C_1(\cosh \kappa x - \cos \kappa x) + C_2(\sinh \kappa x - \sin \kappa x), \qquad (3.48)$$

where $C_1$, $C_2$ are constants. The boundary conditions at $x = L$ follow from dynamic equilibrium of the attached mass: lateral force acting on the mass from the post equals mass times acceleration at $x = L$, and rotation moment acting on the mass from the post equals the moment of inertia times rotational acceleration at $x = L$. According to Eqs. (3.43), (3.44), and the sign convention of Fig. 3.22, these boundary conditions are as follows:

$$EIu''(L) = J\omega^2 u'(L),$$
$$EIu'''(L) = -m\omega^2 u(L). \tag{3.49}$$

Substitution of (3.48) into boundary conditions (3.49) leads to a homogeneous system of two linear equations for $C_1$, $C_2$. For non-zero solutions of this system to exist, the determinant must be equal to zero. After straightforward but extensive algebra (see [17] for effective ways of performing such calculations), this results in the following equation for the non-dimensional frequency variable $\beta = \kappa L$:

$$1 + \cosh\beta\cos\beta - m_R\beta(\cosh\beta\sin\beta - \sinh\beta\cos\beta)$$
$$- J_R\beta^3(\cosh\beta\sin\beta + \sinh\beta\cos\beta) \tag{3.50}$$
$$+ m_R J_R\beta^4(1 - \cosh\beta\cos\beta) = 0.$$

Here, $m_R$ and $J_R$ are non-dimensional parameters describing inertial properties of the payload relative to those of the post:

$$m_R = \frac{m}{\rho AL}, \quad J_R = \frac{J}{\rho AL^3}. \tag{3.51}$$

Equation (3.50) must be solved numerically. In our case, $m = 0.2$ kg, $\rho = 7.8 \cdot 10^3$ kg/m³, $A = \pi d^2/4 = 4.91 \cdot 10^{-4}$ m², $L = 0.15$ m, $J = 2 \cdot 10^{-4}$ kg·m²; $m_R = 0.348$; $J_R = 0.0155$; the first two roots are $\beta_1 \approx 1.483$ and $\beta_2 \approx 3.479$. Comparing that to the CF (clamped-free) case of Fig. 3.23, we notice that the payload appendage leads to substantial reduction in natural frequencies. Natural frequencies in Hz equal

$$f_n = \frac{1}{2\pi}\frac{\beta_n^2}{L^2}\sqrt{\frac{EI}{\rho A}}. \tag{3.52}$$

In our case, $E = 2 \cdot 10^{11}$ N/m², $I = \pi d^4/64 = 1.917 \cdot 10^{-8}$ m⁴. This gives for the first natural frequency $f_1 = 492$ Hz, for the second,

$f_2 = 2709$ Hz. The deformed central axis is described by Eq. (3.48) where the ratio of constants is given by

$$\frac{C_2}{C_1} = -\frac{(\cosh \beta + \cos \beta) - \beta^3 J_R(\sinh \beta + \sin \beta)}{(\sinh \beta + \sin \beta) - \beta^3 J_R(\cosh \beta - \cos \beta)}$$
$$= \frac{(\sinh \beta - \sin \beta) + \beta m_R(\cosh \beta - \cos \beta)}{(\cosh \beta + \cos \beta) + \beta m_R(\sinh \beta - \sin \beta)}. \tag{3.53}$$

For the first mode, calculations give $C_2/C_1 = -0.756$; for the second, $C_2/C_1 = -1.057$. The first and second mode shapes are shown in Figs. 3.24(b) and 3.24(c).

Detailed practical guidance for analysis of elastic beams and frames can be found in the handbook by Karnovsky and Lebed [17].

**Example 3.3. Dynamic deflection of the optical post system.**
Suppose the foundation supporting the optical post system described in Example 3.2 experiences horizontal displacement $u_0$ and angular displacement $\theta_0$. Dynamic deflection and rotation of the payload will be related to the deflection and rotation of the foundation, in terms of complex amplitudes, via a $2 \times 2$ matrix of frequency-dependent transfer functions:

$$\begin{pmatrix} u_L \\ u'_L \end{pmatrix} = \begin{pmatrix} T_{11}(\omega) & T_{12}(\omega) \\ T_{21}(\omega) & T_{22}(\omega) \end{pmatrix} \begin{pmatrix} u_0 \\ u'_0 \end{pmatrix},$$

where $u_L$, $u'_L$ are the displacement and the slope at the upper end of the post $(x = L)$; $u'_0 = \theta_0$.

For simpler calculations, consider the effect of linear excitation $u_0$ and the effect of rotational excitation $\theta_0$ separately. In case of linear excitation, zero boundary conditions at $x = 0$ are replaced by inhomogeneous boundary conditions

$$u(0) = u_0, \, u'(0) = 0, \tag{3.54}$$

that should be considered in conjunction with the boundary conditions at $x = L$ (3.49). The following set of special functions, called Krylov functions, greatly simplifies dynamical calculations for beams and frames:

$$S(x) = \frac{1}{2}(\cosh x + \cos x); T(x) = \frac{1}{2}(\sinh x + \sin x);$$

$$U(x) = \frac{1}{2}(\cosh x - \cos x); V(x) = \frac{1}{2}(\sinh x - \sin x). \tag{3.55}$$

A remarkable property of Krylov functions is the circular relationship of their derivatives:

$$S'(x) = V(x); V'(x) = U(x); U'(x) = T(x); T'(x) = S(x).$$

This leads to the "unity matrix" property:

$$S(0) = 1; T(0) = 0; U(0) = 0; V(0) = 0.$$
$$S'(0) = 0; T'(0) = 1; U'(0) = 0; V'(0) = 0.$$
$$S''(0) = 0; T''(0) = 0; U''(0) = 1; V''(0) = 0.$$
$$S'''(0) = 0; T'''(0) = 0; U'''(0) = 0; V'''(0) = 1.$$

Further properties of Krylov functions can be found in the handbook [17]. The general solution of Eq. (3.46) is

$$u(x) = C_1 S(\kappa x) + C_2 T(\kappa x) + C_3 U(\kappa x) + C_4 V(\kappa x).$$

Due to the boundary conditions (3.54) and the "unity matrix" property of Krylov functions, this solution takes the form

$$u(x) = u_0 S(\kappa x) + C_3 U(\kappa x) + C_4 V(\kappa x).$$

Boundary conditions at $x = L$ (3.49) lead to the system of two linear equations

$$EI\kappa^2[u_0 U(\kappa L) + C_3 S(\kappa L) + C_4 T(\kappa L)]$$
$$= \omega^2 J\kappa[u_0 V(\kappa L) + C_3 T(\kappa L) + C_4 U(\kappa L)],$$
$$EI\kappa^3[u_0 T(\kappa L) + C_3 V(\kappa L) + C_4 S(\kappa L)]$$
$$= -\omega^2 m[u_0 S(\kappa L) + C_3 U(\kappa L) + C_4 V(\kappa L)],$$

that can be resolved for $C_3$ and $C_4$ in terms of $u_0$. After considerable but straightforward algebra we obtain the following formulas for transfer functions $T_{11}$ and $T_{21}$:

$$T_{11} = \frac{u(L)}{u(0)} = \frac{S(\kappa L) - (\kappa L)^3 J_R T(\kappa L)}{\det(\kappa L)};$$

$$T_{21} = \frac{u'(L)}{u(0)} = \frac{\kappa[V(\kappa L) + (\kappa L)m_R U(\kappa L)]}{\det(\kappa L)},$$

where

$$\det(\beta) = \frac{1}{2}[1 + \cosh\beta\cos\beta - m_R\beta(\cosh\beta\sin\beta - \sinh\beta\cos\beta)$$
$$- J_R\beta^3(\cosh\beta\sin\beta + \sinh\beta\cos\beta) + m_R J_R\beta^4(1 - \cosh\beta\cos\beta)],$$

$\beta = \kappa L$, $\kappa$ is the wave number defined by Eq. (3.47), $m_R$ and $J_R$ are non-dimensional parameters (3.51).

The case of rotational excitation is solved in a similar way. The inhomogeneous boundary conditions are

$$u(0) = 0, u'(0) = \theta_0.$$

The general solution of (3.46) satisfying these conditions is

$$u(x) = \frac{1}{\kappa}\theta_0 T(\kappa x) + C_1 U(\kappa x) + C_2 V(\kappa x).$$

Substituting this into boundary conditions (3.49) and resolving for $C_1$, $C_2$, one obtains, after some algebra,

$$T_{12} = \frac{u(L)}{u'(0)} = \frac{T(\kappa L) - (\kappa L)^3 J_R U(\kappa L)}{\kappa \det(\kappa L)};$$

$$T_{22} = \frac{u'(L)}{u'(0)} = \frac{[S(\kappa L) + (\kappa L)m_R V(\kappa L)]}{\det(\kappa L)}.$$

Figure 3.25 shows the amplitudes of the transfer functions. To avoid infinite amplitudes, small damping has been assumed by introducing a complex elastic modulus $E_c = E(1+i\eta)$, $\eta = 0.01$. The graph shows that at low frequencies the post behaves as a rigid body, which is the desired behavior. However, as the excitation frequency approaches the lowest natural frequency of the loaded post, deviations from the desired position and orientation of the attached optomechanical payload can be very significant. We'll return to this model below in Example 3.4.

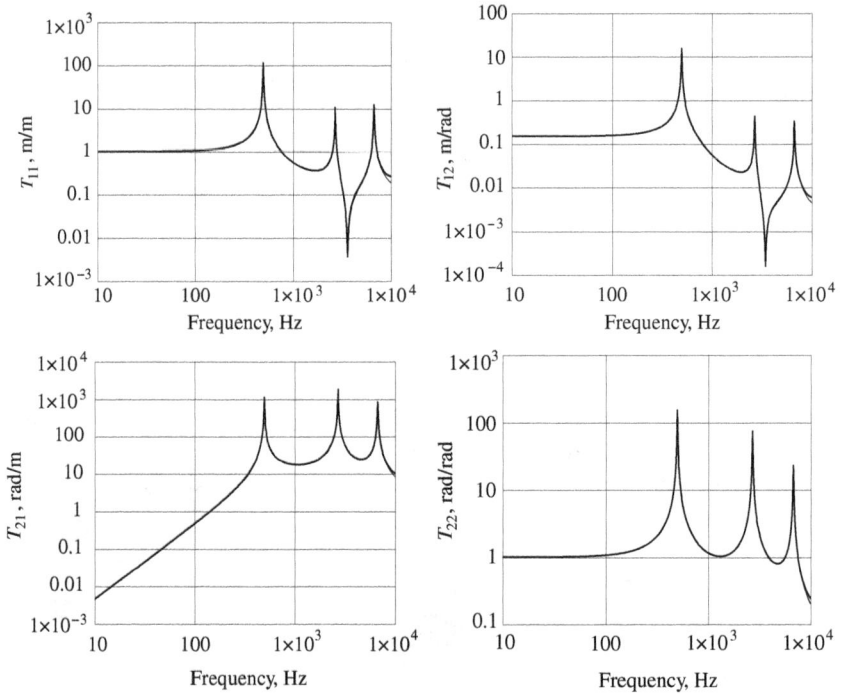

**Figure 3.25.** Transfer functions of linear and angular motion from the base to the top of the optical post in Example 3.3. Thick lines — direct calculation, thin lines — sums of modal contributions of the first three modes (see Example 3.4).

Modal analysis is a powerful tool for studying dynamic behavior of mechanical and optomechanical systems using continuous models. A continuous system has, generally, an infinite number of natural modes, and its reaction to a distributed force $p(x)$ acting at a frequency $\omega$ is a sum of contributions from all the modes, analogous (assuming uniform damping) to Eq. (3.36):

$$u(x) = \sum_{k=1}^{\infty} \frac{\int p(x)\varphi_k(x)dx}{\omega_k^2(1+i\eta) - \omega^2}\varphi_k(x), \qquad (3.56)$$

where $x$ represents spatial coordinates, and $p(x)$ is the corresponding force density such as surface pressure. Natural modes $\varphi_k$ are normalized with respect to kinetic energy

$$\int \rho(x)[\varphi_k(x)]^2 dx = 1. \qquad (3.57)$$

For non-uniform damping, a realistic first approximation can be achieved by assigning specific loss factors, $\eta_k$, to each mode. Calculation of these loss factors is often included in standard finite-element routines. In case of a concentrated force $F(x_0)$, Eq. (3.56) takes the form

$$u(x) = F(x_0) \sum_{k=1}^{\infty} \frac{\varphi_k(x_0)\varphi_k(x)}{\omega_k^2(1 + i\eta_k) - \omega^2}. \tag{3.58}$$

If $x = x_0$, Eq. (3.58) gives the following expression for local dynamic compliance:

$$C(x,\omega) = \frac{u(x)}{F(x)} = \sum_{k=1}^{\infty} \frac{1}{m_k(x)} \frac{1}{\omega_k^2(1 + i\eta_k) - \omega^2}, \tag{3.59}$$

where each term under the sum has the same form as a dynamic compliance of a SDOF oscillator (see Exercise 3.5), and $m_k(x) = 1/[\varphi_k(x)]^2$ can be interpreted as an equivalent modal mass. Recalling Section 3.3, one obtains the following estimate of the RMS displacement modal response caused by the random input in the vicinity of the $k$th natural frequency, as defined by the given power spectral density of the input force $W_F(f)$:

$$[u_{\mathrm{RMS}}(x)]_k = \frac{\sqrt{W_F(f_k)}}{\sqrt{32\pi^3 f_k^3 \eta_k}} \frac{1}{m_k(x)}, \tag{3.60}$$

where $f_k = \omega_k/2\pi$ are natural frequencies in Hz. Note that formulas (3.59) and (3.60) are analogous to Eqs. (3.37) and (3.38) for discrete MDOF models considered in the previous section. Mean square of the total response is closely approximated, at least for small damping, by the sum of modal contributions [12]:

$$\langle u(x)^2 \rangle = \sum_{k=1}^{\infty} \frac{W_F(f_k)}{32\pi^3 f_k^3 \eta_k} \frac{1}{[m_k(x)]^2}.$$

**Example 3.4. Forced vibration of the optical post as a sum of contributions of natural modes.** As in the previous example, let us consider the case when the foundation supporting the optical post system experiences horizontal displacement $u_0$ and angular displacement $\theta_0$, and find the resulting motion of the optical appendage at the top of the post as the sum of the contribution of resonance modes, following the approach based on Eq. (3.56). To find the motion under the excitation from the foundation, we can, the same way we did in the previous section when deriving Eq. (3.30), consider the problem in the coordinate system moving (in linear motion and rotation) with the foundation, and solve for relative motion while applying inertia forces. In this case the inertia loads consist of distributed and concentrated transversal load and a concentrated moment:

$$p_{\text{inertia}} = -[\rho A + m\delta(x - L)](\ddot{u}_0 + x\ddot{\theta}_0), \qquad (3.61)$$

$$M_{\text{inertia}} = J\ddot{\theta}_0\delta(x - L). \qquad (3.62)$$

Here $\delta(x)$ is the Kronecker delta function describing a concentrated force; see Example 3.2 for other notations.

The natural modes had been found in Example 3.2. From Eqs. (3.48) and (3.53) we can write, using Krylov functions (3.55), the following compact expression for the (non-normalized) natural modes:

$$\varphi_k(x) = [S(\beta_k) + m_R\beta_k V(\beta_k)]U\left(\frac{\beta_k}{L}x\right)$$
$$- [V(\beta_k) + m_R\beta_k U(\beta_k)]V\left(\frac{\beta_k}{L}x\right), \qquad (3.63)$$

where $\beta_k$ — solutions of the characteristic equation (3.50): $\beta_1 \approx 1.483$, $\beta_2 \approx 3.479$, $\beta_3 \approx 5.496$. These modes must be normalized with respect to the total kinetic energy form of the system (including the beam and the appendage) before using them in the modal expansion (3.56).

Again, consider the effects of linear excitation, $u_0$, and angular excitation, $\theta_0$, separately. For the linear excitation $u_0$ alone, the complex amplitude of relative motion is, from (3.56) and (3.61),

$$u_{\text{rel}}(x) = u_0 \sum_k \frac{\omega^2[\int_0^L \rho A \varphi_k(x)dx + m\varphi_k(L)]}{\int_0^L \rho A \varphi_k^2(x)dx + m(\varphi_k(L))^2 + J(\varphi_k'(L))^2}$$
$$\times \frac{\varphi_k(x)}{\omega_k^2(1 + i\eta) - \omega^2}.$$

Absolute motion $u = u_{\text{rel}} + u_0$. Since, by assumption, $\rho A = \text{const}$, we have, using non-dimensional parameters (3.51),

$$\frac{u(x)}{u_0} = 1 + \omega^2 \sum_k \frac{\int_0^L \varphi_k(x)dx + m_R L \varphi_K(L)}{\int_0^L \varphi_k^2(x)dx + m_R L(\varphi_K(L))^2 + J_R L^3(\varphi_k'(L))^2}$$

$$\times \frac{\varphi_k(x)}{\omega_k^2(1 + i\eta) - \omega^2}.$$

Hence, for the transfer functions,

$$T_{11}(\omega) = \frac{u(L)}{u(0)} = 1 + \omega^2 \sum_k \frac{z_k^{(11)}}{\omega_k^2(1 + i\eta) - \omega^2},$$

$$T_{21}(\omega) = \frac{u'(L)}{u(0)} = \omega^2 \sum_k \frac{z_k^{(21)}}{\omega_k^2(1 + i\eta) - \omega^2},$$

$$(3.64)$$

where

$$z_k^{(11)} = \frac{(\int_0^L \varphi_k(x)dx + m_R L \varphi_k(L))\varphi_k(L)}{N_k},$$

$$z_k^{(21)} = \frac{(\int_0^L \varphi_k(x)dx + m_R L \varphi_k(L))\varphi_k'(L)}{N_k},$$

$$(3.65)$$

$$N_k = \int_0^L \varphi_k^2(x)dx + m_R L(\varphi_k(L))^2 + J_R L^3(\varphi_k'(L))^2.$$

Integrals of normal mode functions $\varphi_k(x)$ (3.63) in the equations above can be calculated analytically, using the properties of Krylov functions, or numerically. Their numerical values, as well as other constants present in Eqs. (3.64)–(3.65), are summarized below in Table 3.3.

To find the reaction to rotational excitation $u'(0) = \theta_0$, we use inertial forces (3.61), (3.62), associated with rotation $\theta_0$, in Eq. (3.56) to obtain, for the complex amplitude of relative displacement,

$$u_{\text{rel}}(x) = \theta_0 \sum_k \frac{\omega^2[\int_0^L \rho A x \varphi_k(x)dx + mL\varphi_k(L) + J\varphi_k'(L)]}{\int_0^L \rho A \varphi_k^2(x)dx + m(\varphi_k(L))^2 + J(\varphi_k'(L))^2}$$

$$\times \frac{\varphi_k(x)}{\omega_k^2(1 + i\eta) - \omega^2}.$$

**Table 3.3.** Numerical values of parameters (to Example 3.4).

| Parameter | Mode No. | | |
|---|---|---|---|
| | $k = 1$ | $k = 2$ | $k = 3$ |
| $\varphi_k(L)$ | 1.041 | $-2.677$ | $-43.159$ |
| $\varphi_k'(L)$ (m$^{-1}$) | 10.279 | $-468.497$ | $2.971 \cdot 10^3$ |
| $\int_0^L \varphi_k(x)dx$ (m) | 0.0592 | 0.9444 | 7.0894 |
| $\int_0^L x\varphi_k(x)dx$ (m$^2$) | $6.492 \cdot 10^{-3}$ | 0.0783 | 0.316 |
| $\int_0^L \varphi_k^2(x)dx$ (m) | 0.0388 | 8.5314 | $1.0065 \cdot 10^3$ |
| $z_k^{(11)}$ | 1.171 | $-0.106$ | $-0.133$ |
| $z_k^{(12)}$ (m/rad) | 0.157 | $-4.312 \cdot 10^{-3}$ | $-3.668 \cdot 10^{-3}$ |
| $z_k^{(21)}$ (rad/m) | 11.561 | $-18.504$ | 9.180 |
| $z_k^{(22)}$ | 1.546 | $-0.755$ | 0.252 |
| $N_k$, (m) | 0.101 | 20.371 | $1.565 \cdot 10^3$ |

Following the same procedure as for linear excitation, one arrives at the equation for the absolute displacement:

$$\frac{u(x)}{\theta_0} = x + \omega^2 \sum_k \frac{\int_0^L x\varphi_k(x)dx + m_R L^2 \varphi_k(L) + J_R L^3 (\varphi_k'(L))}{\int_0^L \varphi_k^2(x)dx + m_R L(\varphi_k(L))^2 + J_R L^3 (\varphi_k'(L))^2}$$
$$\times \frac{\varphi_k(x)}{\omega_k^2(1 + i\eta) - \omega^2}.$$

Transfer functions from the rotation of the base to displacement and rotation at the top of the post are given by the following formulas:

$$T_{12}(\omega) = \frac{u(L)}{u'(0)} = L + \omega^2 \sum_k \frac{z_k^{(12)}}{\omega_k^2(1 + i\eta) - \omega^2},$$

$$T_{22}(\omega) = \frac{u'(L)}{u'(0)} = 1 + \omega^2 \sum_k \frac{z_k^{(22)}}{\omega_k^2(1 + i\eta) - \omega^2},$$

(3.66)

where

$$z_k^{(12)} = \frac{(\int_0^L x\varphi_k(x)dx + m_R L^2 \varphi_k(L) + J_R L^3 (\varphi_k')(L))\varphi_k(L)}{N_k},$$

$$z_k^{(22)} = \frac{(\int_0^L x\varphi_k(x)dx + m_R L^2 \varphi_k(L) + J_R L^3 (\varphi_k')(L))\varphi_k'(L)}{N_k}.$$

(3.67)

Note that the first (constant) terms in the expressions for transfer functions (3.64), (3.66) reflect the motions of the post as a rigid body together with the foundation. The constants present in (3.64)–(3.67) can be found in Table 3.3.

Figure 3.25 shows that the modal approximation of transfer functions using the first three modes gives a practically exact solution in the frequency range encompassing the first three natural frequencies.

Finally, let us estimate the resonance reaction of the optical payload to the random excitation of the base. Consider the linear excitation, $u_0$. Assume $W_{\ddot{u}_0}(f) = 10^{-9}\, g^2/\text{Hz}$ as a realistic estimate of the PSD of the horizontal acceleration of the base (see Chapter 5). The same type of calculation that led to Eq. (3.24) (see also (3.59) and (3.60)) allows one to estimate the contribution of the resonance at $f = f_k$ to the RMS values of linear and rotational motion of the optical payload. Since $\sqrt{W_{\ddot{u}_0}(f)} = (2\pi f)^2 \sqrt{W_{u_0}(f)}$ we have, using the expression for the transfer function $T_{11}$ (3.64), the formula for the contribution of the first resonance to $u_{\text{RMS}}$:

$$[u_{\text{RMS}}(L)]_1 = \frac{\sqrt{W_{\ddot{u}_0}(f_1)}\,|z_1^{(11)}|}{\sqrt{32\pi^3 f_1^3 \eta}}.$$

In our case $f_1 = 492$ Hz (see Example 3.2), $\eta = 0.01$ (see Example 3.3), so the calculation gives

$$[u_{\text{RMS}}(L)]_1 = \frac{9.8\ \text{ms}^{-2}\sqrt{10^{-9}\ \text{s}} \cdot 1.171}{\sqrt{32\pi^3 \cdot 492^3\ \text{s}^{-3} \cdot 0.01}} \approx 1.1 \cdot 10^{-8}\ \text{m}.$$

Rotational displacement is estimated the same way using the transfer function $T_{21}$ (3.64):

$$[u'_{\text{RMS}}(L)]_1 = \frac{9.8\ \text{ms}^{-2}\sqrt{10^{-9}\ \text{s}} \cdot 11.561\ \text{rad/m}}{\sqrt{32\pi^3 \cdot 492^3\ \text{s}^{-3} \cdot 0.01}} \approx 1.0 \cdot 10^{-7}\ \text{rad}.$$

Contributions from higher modes are an order of magnitude smaller.

It should be noted that Examples 3.2–3.4 consider an idealized model of the optical post that disregards important factors such as finite linear and rotational stiffness of the attachment surface and the post-payload interface, as well as dynamic properties of the appendage itself. Experimental studies and real-life dynamic response functions of posts and other optomechanical elements are presented below in Chapter 8.

**Example 3.5. Simplified model of a flexible vibration-isolated platform.** Vibration isolation system consists of two main components: soft isolators and a stiff platform. Detailed description of isolators and platforms will be given below in Chapter 6. This example is a model study that clarifies principal aspects of dynamic behavior of such systems. This example was originally developed as part of Newport Corporation's technical note.

The dynamics of a rigid body supported by springs had been analyzed above in Section 3.4 and Exercise 3.8. The rigid body has six degrees of freedom and six natural modes corresponding to translational and rotational "bouncing" on isolators. Real platforms are not absolutely rigid. They are continuous systems that have a sequence of natural frequencies and natural modes describing flexural vibrations (bending and twisting) of the structure. These frequencies are usually much higher than the frequencies of the rigid body oscillating on isolators. The simplest model that reflects these dynamic properties is depicted in Fig. 3.26. Here the isolated platform is represented by a beam simply supported on two springs. Motion is restricted to the vertical direction in the plane of the drawing. We'll consider two cases of excitation: linear motion of the foundation with the displacement amplitude $u_0$, Fig. 3.26(a), modeling the vibration transmissibility from the floor, and force excitation with amplitude $F$, Fig. 3.26(b), modeling effects of acoustical, atmospheric or on-board sources of vibration. Applying forces and elastic supports at the ends of the beam allows for simpler analytical solutions obtained just by using proper boundary conditions in each case.

Consider first the free vibrations of the beam. They are governed, just like in previous examples of the optical post, by Eqs. (3.46) and (3.47).

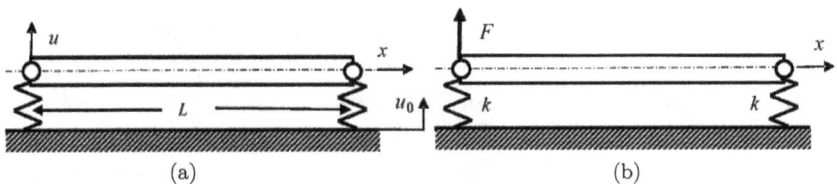

**Figure 3.26.** Model of isolated platform (Example 3.5): (a) kinematic excitation; (b) force excitation.

Boundary conditions state that the rotation moments (3.43) are zeroes at both ends, therefore,

$$u''(0) = 0; u''(L) = 0. \qquad (3.68)$$

Elastic support conditions at both ends, according to Eqs. (3.43), (3.44) and sign convention of Fig. 3.22, take the following form:

$$EIu'''(0) = -ku(0); EIu'''(L) = ku(L), \qquad (3.69)$$

where $L$ is the length of the beam, $EI$ is the bending stiffness, and $k$ is the spring rate.

Before proceeding to the complete solution, let us consider two partial models of this system: a rigid beam oscillating on springs and a free elastic beam without supports.

A rigid beam, under the stated restrictions, is described by two generalized coordinates (degrees of freedom): vertical displacement of its center of gravity, $u_c$, and rotation around the center of gravity, $\theta$, as shown in Fig. 3.27, so that

$$u(x) = u_c + \left(x - \frac{L}{2}\right)\theta. \qquad (3.70)$$

Using the methodology introduced in the previous section for the rigid body model, the expressions for kinetic energy,

$$T_K = \frac{1}{2}m\dot{u}_c^2 + \frac{1}{2}J\dot{\theta}^2,$$

and potential energy,

$$U = \frac{1}{2}k\left[\left(u_c - \frac{L}{2}\theta\right)^2 + \left(u_c + \frac{L}{2}\theta\right)^2\right] = k\left[u_c^2 + \left(\frac{L}{2}\theta\right)^2\right],$$

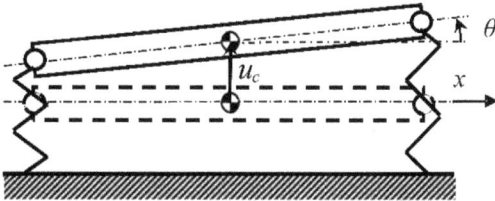

**Figure 3.27.** Motion of the beam as a rigid body.

lead to the following diagonal mass and stiffness matrices:

$$\mathbf{M} = \begin{pmatrix} m & 0 \\ 0 & J \end{pmatrix}, \quad \mathbf{C} = \begin{pmatrix} 2k & 0 \\ 0 & k\frac{L^2}{2} \end{pmatrix}.$$

The characteristic equation (3.32) gives two natural frequencies: first, corresponding to the translational motion,

$$\omega_{r1} = \sqrt{\frac{2k}{m}}, \quad f_{r1} = \frac{1}{2\pi}\sqrt{\frac{2k}{m}}, \tag{3.71}$$

and second, corresponding to rotation around the center,

$$\omega_{r2} = \sqrt{\frac{kL^2}{2J}}, \quad f_{r2} = \frac{1}{2\pi}\sqrt{\frac{kL^2}{2J}}. \tag{3.72}$$

For thin beams, the moment of inertia $J$ is, approximately,

$$J = \frac{mL^2}{12}. \tag{3.73}$$

Equations (3.71)–(3.73) lead to the relationship $f_{r2} = f_{r1}\sqrt{3}$.

State-of-the-art vibration isolators can produce the natural frequency $f_0$ of vertical vibration relatively independent of the load. To reflect this, we assume in subsequent calculations a fixed value $f_0 = 1$. That gives $k = m \cdot (2\pi f_0)^2/2$. We'll conduct all calculations in non-dimensional form. Numerical results can be obtained by substituting specific values for $f_0$, $m$, and $L$.

In case of kinematic excitation $u_0$, Fig. 3.26(a), the rotational mode is not excited because of symmetry, and the rigid beam behaves as an SDOF system. Transmissibility, $u/u_0$, is shown in Fig. 3.28 by a thin line. To avoid infinite amplitudes at the resonances, we introduced frequency-independent damping by using, in place of $k$, complex stiffness $k(1+i\eta_0)$ with the loss factor $\eta_0 = 0.2$ (see Section 3.2).

In case of force excitation, Fig. 3.26(b), the force $F$ applied at the end $x = 0$ can be replaced by the equivalent system of the force $F$

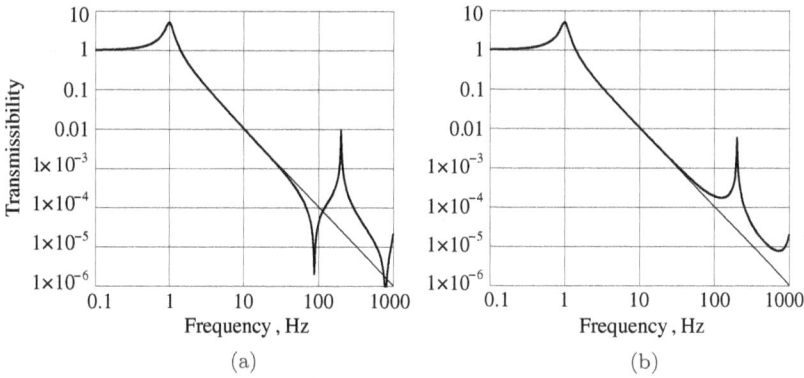

**Figure 3.28.** Example 3.5: Transmissibilities of the isolated platform to the edge of the beam (a) and the center of the beam (b). Thick line — full model, thin line — rigid beam. Only symmetric modes are excited by symmetric kinematic excitation (Fig. 3.26(a)).

applied at the center and the moment $-FL/2$. So,

$$\begin{pmatrix} u_c \\ \theta \end{pmatrix} = (\mathbf{C} - \omega^2 \mathbf{M})^{-1} \begin{pmatrix} F \\ M \end{pmatrix} = \begin{pmatrix} \frac{1}{2k - m\omega^2} & 0 \\ 0 & \frac{1}{\frac{1}{2}kL^2 - J\omega^2} \end{pmatrix} \begin{pmatrix} F \\ -\frac{FL}{2} \end{pmatrix}.$$

According to Eq. (3.70), the displacement $u(x)$ is defined by

$$u(x) = F \left[ \frac{1}{2k - m\omega^2} + \frac{\frac{L}{2} \left( \frac{L}{2} - x \right)}{k\frac{L^2}{2} - J\omega^2} \right].$$

Recalling (3.73) and substituting

$$2k = m\omega_{r1}^2,$$

gives the expression

$$\frac{u(x)}{F} = \frac{1}{m\omega_{r1}^2} \left( \frac{1}{1 - \frac{\omega^2}{\omega_{r1}^2}} + \frac{L - 2x}{L} \frac{1}{1 - \frac{\omega^2}{3\omega_{r1}^2}} \right). \tag{3.74}$$

This expression, taken at $x = 0$, gives the dynamic compliance of the rigid beam at its end. This function is plotted by thin lines in Fig. 3.29 for $x = 0$ and $x = L/2$. Here, again, we introduced frequency-independent damping by using, in place of $k$, complex stiffness $k(1 + i\eta_0)$ with the loss factor $\eta_0 = 0.2$, and corresponding

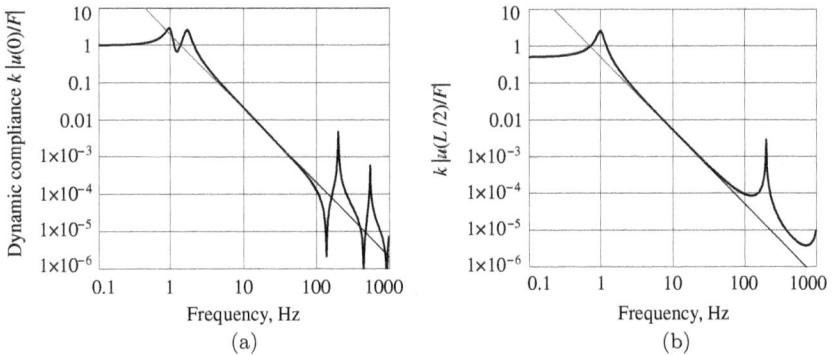

**Figure 3.29.** Example 3.5: Dynamic compliance at the end of the beam (a) and transfer function from the force at one end of the beam to the displacement at the middle of the beam (b). Thick line — beam supported by springs, thin line — rigid beam, dim line — ideal free rigid body. All functions are referred to the static compliance, $1/k$.

complex natural frequencies (see Section 3.2). Note that the rocking mode of the beam is excited in this case along with the linear vertical motion.

Another partial model is the free-floating beam. The modal shape of the vibrating beam is described by Eq. (3.46). If the springs are absent, $k = 0$, we have the classic "free-free" beam of Fig. 3.23. In this case, (3.69) is replaced by

$$u'''(0) = 0; u'''(L) = 0. \tag{3.75}$$

The general solution of (3.46) in terms of Krylov functions (3.55) is

$$u(x) = C_1 S(\kappa x) + C_2 T(\kappa x) + C_3 U(\kappa x) + C_4 V(\kappa x). \tag{3.76}$$

From boundary conditions (3.68), (3.70) at $x = 0$, $C_3 = C_4 = 0$, so

$$u(x) = C_1 S(\kappa x) + C_2 T(\kappa x).$$

where $\kappa$ is the wave number defined by Eq. (3.47). Boundary conditions (3.68), (3.69) at $x = L$ give, using the circular relationship of the derivatives of Krylov functions (Example 3.3),

$$C_1 U(\kappa L) + C_2 V(\kappa L) = 0,$$
$$C_1 T(\kappa L) + C_2 U(\kappa L) = 0.$$

Solution exists if the determinant equals zero, that is, the frequency satisfies the following characteristic equation for the non-dimensional frequency parameter $\beta = \kappa L$:

$$\det(\beta) = U^2(\beta) - T(\beta)V(\beta) = 0, \qquad (3.77)$$

or, equivalently, $\cosh \beta \cos \beta - 1 = 0$.

The solutions, besides the trivial solution $\beta = 0$, are: $\beta_1 \approx 4.730$, $\beta_2 \approx 7.853$; $\beta_3 \approx 10.996$; subsequent roots closely follow $\beta_n = (2n+1)\pi/2$. The non-dimensional parameter $\beta$ defines the frequency by Eq. (3.52). Let us assume the first natural frequency of flexural vibration of the platform $f_1 = 200$ Hz which is a typical value for laboratory size optical tables. According to Eq. (3.52),

$$f_1 = \frac{1}{2\pi} \frac{\beta_1^2}{L^2} \sqrt{\frac{EI}{\rho A}},$$

where $\rho A = m/L$, and the bending stiffness of the beam must assume the value $EI = mL^3(2\pi f_1)^2/\beta_1^4$. The characteristic equation (3.77) is illustrated by the thin line in Fig. 3.30.

Now consider the full model of the elastic beam supported by springs. The modal shape of the beam satisfies Eq. (3.46) and boundary conditions (3.68), (3.69). The general solution is given by Eq. (3.76); boundary conditions at $x = 0$ lead to

$$C_3 = 0, C_4 = -\frac{k}{\kappa^3 EI}C_1,$$

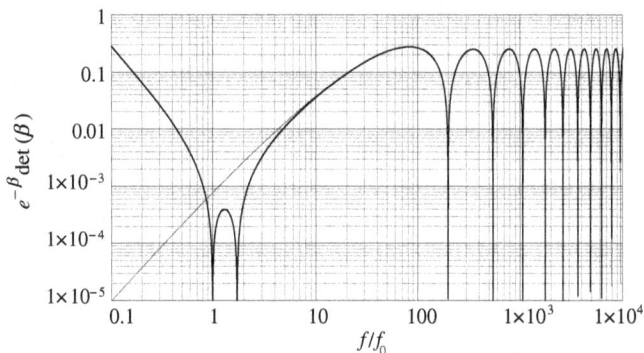

**Figure 3.30.** Example 3.5: Graphical solution of the characteristic equation. Thin line — free-free beam, thick line — beam supported by soft springs.

so that

$$u(x) = C_1 \left( S(\kappa x) - \frac{k}{\kappa^3 EI} V(\kappa x) \right) + C_2 T(\kappa x).$$

Boundary conditions at $x = L$ give the system of two equations for $C_1, C_2$:

$$C_1 \left( U(\beta) - \frac{k_r}{\beta^3} T(\beta) \right) + C_2 V(\beta) = 0,$$

$$C_1 \left[ T(\beta) - \frac{2k_r}{\beta^3} S(\beta) + \left( \frac{k_r}{\beta^3} \right)^2 V(\beta) \right] + C_2 \left( U(\beta) - \frac{k_r}{\beta^3} T(\beta) \right) = 0.$$

Here $\beta = \kappa L$, and $k_r$ is a non-dimensional parameter characterizing the stiffness of the spring:

$$k_r = \frac{kL^3}{EI}.$$

In this example $k = m(2\pi f_0)^2/2$, $EI = mL^3(2\pi f_1)^2/\beta_1^4$, so $k_r = (f_0/f_1)^2 \beta_1^4/2 = (1 \text{ Hz}/200 \text{ Hz})^2 \cdot 4.730^4/2 \approx 6.257 \cdot 10^{-3}$.

The characteristic equation is

$$\det(\beta) = \left( U(\beta) - \frac{k_r}{\beta^3} T(\beta) \right)^2 - \left[ T(\beta) - \frac{2k_r}{\beta^3} S(\beta) + \left( \frac{k_r}{\beta^3} \right)^2 V(\beta) \right] V(\beta) = 0.$$

This equation is illustrated by the thick line in Fig. 3.30. The solutions are almost identical (within 0.01 Hz in terms of frequency) with that of Eq. (3.77), but two additional solutions appear, corresponding to the natural frequencies of the beam as a rigid body, practically coincident with $f_{r1}$ and $f_{r2}$.

Figure 3.31 shows two natural modes corresponding to rigid motion of the beam and three lowest natural modes of flexural motion of the beam. The two groups of modes are practically independent of each other, divided by a large frequency interval where the platform behaves essentially as a free-floating rigid body. We'll illustrate it by studying forced vibration of the system under kinematic excitation, Fig. 3.26(a), and force excitation, Fig. 3.26(b).

In case of kinematic excitation of the foundation with the amplitude $u_0$, the boundary conditions (3.68) are still valid. Boundary

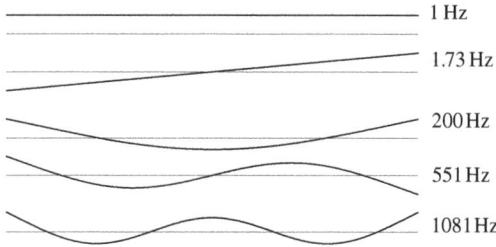

**Figure 3.31.** Example 3.5. Rigid-body and flexural modes of the isolated beam drawn with the same amplitude.

conditions (3.69) take the inhomogeneous form:

$$EIu'''(0) = -k(u(0) - u_0); EIu'''(L) = k(u(L) - u_0).$$

The solution (3.76) satisfying the conditions at $x = 0$ is

$$u(x) = C_1 \left( S(\kappa x) - \frac{k_r}{\beta^3} V(\kappa x) \right) + C_2 T(\kappa x) + \frac{k_r}{\beta^3} V(\kappa x) u_0. \quad (3.78)$$

Boundary conditions at $x = L$ provide the system of two equations for the constants $C_1$, $C_2$:

$$\begin{bmatrix} U(\beta) - \frac{k_r}{\beta^3} T(\beta) & V(\beta) \\ T(\beta) - 2\frac{k_r}{\beta^3} S(\beta) + \left( \frac{k_r}{\beta^3} \right)^2 V(\beta) & U(\beta) - \frac{k_r}{\beta^3} T(\beta) \end{bmatrix} \begin{pmatrix} C_1 \\ C_2 \end{pmatrix}$$

$$= -\frac{k_r}{\beta^3} \begin{bmatrix} T(\beta) \\ S(\beta) + 1 - \frac{k_r}{\beta^3} V(\beta) \end{bmatrix} u_0. \quad (3.79)$$

Resolving (3.79) for $C_1$, $C_2$ and substituting into (3.78), one obtains the expression for transmissibility function $u/u_0$. It is graphed by thick lines in Fig. 3.28 for the end of the beam and the center of the beam. To avoid infinite amplitudes at resonances, damping was introduced into the system, characterized by a loss factor $\eta_0 = 0.2$ for the springs and $\eta = 0.01$ for the beam, by substituting complex values $k(1 + i\eta_0)$ in place of $k$ and $EI(1 + i\eta)$ in place of $EI$.

In case of force excitation, Fig. 3.26(b), the boundary conditions are (3.68) and

$$EIu'''(0) = -ku(0) + F; EIu'''(L) = ku(L).$$

Calculation of complex amplitudes $u(x)$ is completely analogous to the previous case. Figure 3.29 shows dynamic compliance at the edge, as well as transfer functions from the force at one end of the beam to the displacement at the middle of the beam under the same assumptions about the damping.

The analytical solutions allow for an insight into practically important questions about the relative displacement of points of the isolated platform and misalignment of parts of the platform. These factors define the image movement and line-of-sight jitter that may affect optical performance. Figure 3.32 shows the maximum relative displacement, $\Delta u$, of two points of the beam and the maximum misalignment (maximum difference between slopes, $\alpha = u'(x)$, at two points of the beam) in case of kinematic excitation from the floor with the amplitude $u_0$ (Fig. 3.26(a)). To analyze the relative displacement of points of the isolated platform and misalignment of parts of the platform in case of force excitation, Fig. 3.26(b), we need to subtract the rigid-body motion from the total displacement. The resulting flexural relative displacement and misalignment are plotted in Fig. 3.33.

In both cases only very small quasi-static deflections are noticed in the frequency range of "rigid body" behavior of the beam (up to about 150 Hz). These deformations are caused by slowly changing inertia forces. On the other hand, deviations from rigid-body behavior around the resonance frequencies are accompanied by large misalignments and relative flexural displacements of points of the beam. Peak relative displacement at the flexural resonance is about twice the absolute displacement.

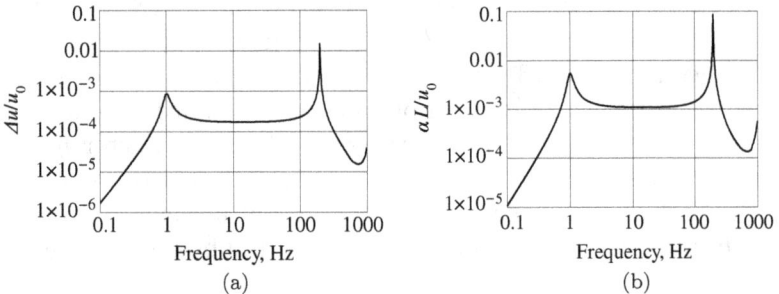

**Figure 3.32.** Example 3.5: (a) Maximum relative displacement between points of the beam, $\Delta u$, and (b) maximum misalignment between two areas of the beam, $\alpha$, caused by kinematic excitation from the foundation, Fig. 3.26(a), per unit displacement of the foundation.

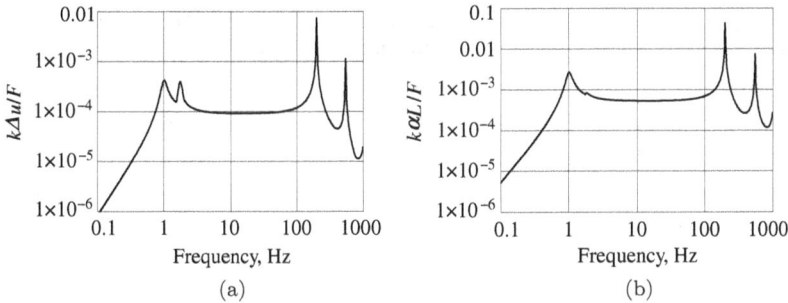

**Figure 3.33.** Example 3.5: (a) Maximum relative flexural displacement between points of the beam, $\Delta u$, and (b) maximum misalignment between two areas of the beam, $\alpha$, caused by force excitation, Fig. 3.26(b), per unit force. All functions are reduced to non-dimensional form.

It should be noted that, although this model example captures the main physical properties of isolated platforms, there is a substantial difference from real isolated platforms such as optical tables. The most significant difference is that a table, being a plate-like structure able to deform in three dimensions, has much denser spectrum of resonance frequencies. Besides, real vibration isolators do not maintain the same stiffness and loss factor at all frequencies. The real-life dynamic properties of isolators and optical tabletops, and the resulting dynamic properties of real-life isolated platforms will be considered in Chapters 6 and 7.

Mathematical models of 2D thin-walled structures (plates and shells) are also well developed. Extensive information on this subject can be found in the fundamental monographs by Leissa [18, 19]. Vibration and wave motion in 3D elastic bodies are studied by the dynamic theory of elasticity; the classic monograph [20] can serve as an introduction to this subject.

Practical optomechanical structures are often too complex for analytical treatment. Numerical finite-element analysis (FEA) is a useful tool for researching dynamic behavior of complex mechanical systems including thin-walled elements, elastic and rigid bodies. It is based on reduction of continuous bodies to discrete MDOF models of high dimensions. Commercial software packages such as MSC Nastran [21] include various routines for analyzing free and forced vibration. These routines can be used for calculating absolute and relative angular and linear displacements of sensitive optical elements directly from given dynamic inputs. The tutorial text [22] contains many useful recommendations for FEA of optomechanical systems.

Numerical modal analysis is restricted to a finite number of modes, usually the first few modes that are the main contributors to the sums (3.35)–(3.37). For illustration, Figs. 3.34 and 3.35 show natural modes, obtained by FEA, of L-shape and T-shape plate-like honeycomb platforms of the type used to host large optomechanical applications. These platforms have a rather dense spectrum of

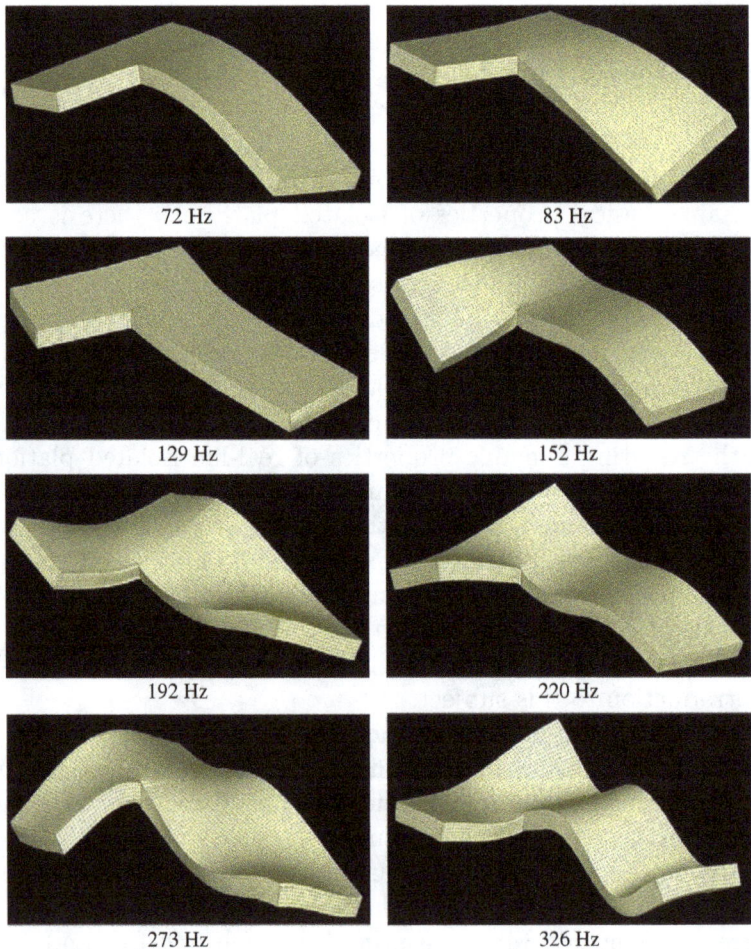

**Figure 3.34.** Natural modes of an L-shaped honeycomb platform comprised of two sections 1.52 m × 3.66 m and 1.22 m × 3.05 m, 0.305 m thick.

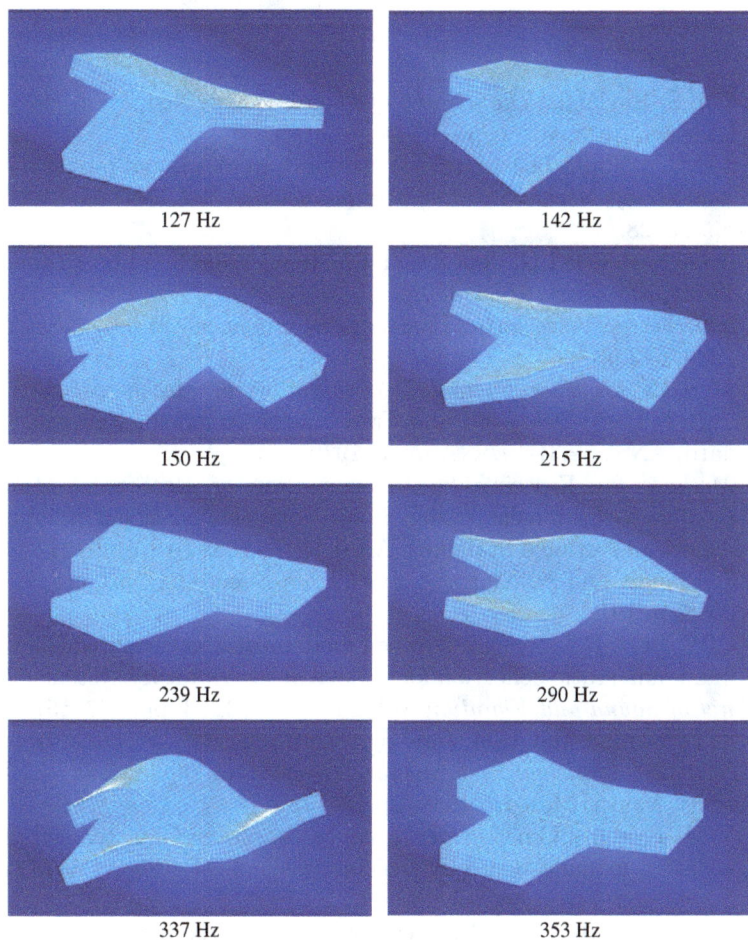

**Figure 3.35.** Natural modes of a T-shaped honeycomb platform comprised of two sections 1.52 m × 3.66 m and 1.52 m × 1.83 m, 0.305 m thick.

natural modes. Most of these modes represent deflection in vertical direction (bending and twisting), although in-plane modes are present, too. Corners of the platform exhibit maximum deflection.

In summary, a wide range of simple and complex mathematical models is available that can be used to understand mechanical properties and transfer functions of optomechanical devices, vibration control platforms, and sensors. The choice of a proper model is a matter of physical intuition and experience.

# References

1. J.W.S. Rayleigh, *The Theory of Sound*, Dover, New York, 1945 and later printings, vol. 1, pp. 93–94.
2. S.H. Crandall, The hysteretic damping model in vibration theory, *Journal of Mechanical Engineering Science*, vol. 205, no. 1, 1991, pp. 23–28.
3. A.D. Nashif, D.I.G. Jones, and J.P. Henderson, *Vibration Damping*, Wiley, New York, 1985, 480 p.
4. J.D. Ferry, *Viscoelastic Properties of Polymers*, Wiley, New York, 1980, 642 p.
5. *Methods for Experimental Determination of Mechanical Mobility.* Part 1: Basic Definition and Transducers. American National Standard, ANSI S2.31-1979 (ASA 31-1979).
6. *Methods for Experimental Determination of Mechanical Mobility.* Part 2: Measurement Using Single-Point Translational Excitation. American National Standard, ANSI S2.32-1982 (ASA 32-1982).
7. J. Blauert and N. Xiang, *Acoustics for Engineers*, Springer, Berlin, 2008, p. 191.
8. C.W. Bert, Relationship between fundamental natural frequency and maximum static deflection for various linear vibratory systems, *Journal of Sound and Vibration*, vol. 162, no. 3, 1993, pp. 547–557.
9. M.D. Genkin and V.M. Ryaboy, *Elastic-inertial Vibration Isolation Systems: Limiting Performance, Optimal Configurations* (in Russian), Nauka, Moscow, 1988, pp. 45–50.
10. V.M. Ryaboy, Limiting performance estimates for active vibration isolation in multi-degree-of-freedom mechanical systems, *Journal of Sound and Vibration*, vol. 186, no. 11, 1995, pp. 1–21.
11. S.H. Crandall and R.B. McCalley Jr., Matrix methods of analysis, C.M. Harris (ed.), in *Shock and Vibration Handbook*, 4th edn., Chapter 28, McGraw-Hill, New York, 1995, and subsequent editions.
12. R.G. DeJong, Statistical methods for analyzing vibrating systems, C.M. Harris (ed.), in *Shock and Vibration Handbook*, 4th edn., McGraw-Hill, New York, 1995 and subsequent editions, pp. 11.11–11.13.
13. H. Himmelblau and S. Rubin, Vibration of a resiliently supported body, C.M. Harris (ed.), in *Shock and Vibration Handbook*, 4th edn., Chapter 3, McGraw-Hill, New York, 1995, and subsequent editions.
14. S.S. Rao, *Mechanical Vibration*, 3rd edn., Addison-Wesley, Reading, MA, 1995, 912 p.

15. S. Timoshenko, D.H. Young, and W. Weaver Jr., *Vibration Problems in Engineering*, 5th edn., Wiley, New York, 1990, 624 p.

16. V.M. Ryaboy, Static and dynamic stability of pneumatic vibration isolators and systems of isolators. *Journal of Sound and Vibration*, vol. 333, 2014, pp. 31–51.

17. I.A. Karnovsky and O.I. Lebed, *Formulas for Structural Dynamics: Tables, Graphs and Solutions*, Chapter 4, McGraw-Hill, New York, 2001, 520 p.

18. A. Leissa, *Vibration of Plates*, Acoustical Society of America, New York, 1993, 354 p.

19. A. Leissa, *Vibration of Shells*, Acoustical Society of America, New York, 1993, 428 p.

20. H. Kolsky, *Stress Waves in Solids*, Dover, New York, 2003, 214 p.

21. A.O. Cifuentes, *Using MSC/NASTRAN: Statics and Dynamics*. Springer, New York, 1989, 472 p.

22. K.B. Doyle, V.L. Genberg, and G.J. Michels, *Integrated Optomechanical Analysis*, 2nd edn., SPIE Press, Bellingham, WA, 2012, 408 p.

## Chapter 4

# Vibration Measurements

We cannot control a phenomenon unless we can measure it. Reliable measurement is the key to successful implementation of vibration control. Inertial sensors — accelerometers and geophones — are the most popular sensors used for precision vibration measurements. The first section of this chapter contains explanations of the principles of action and guidance to specifications of inertial vibration sensors. It also gives a brief review of the relatively new technology for non-contact vibration measurement — laser vibrometry.

It is not enough to collect vibration data: the vibration signal must be processed to provide useful insights into vibration phenomena and transfer functions. Section 4.2 considers the basics of signal processing and the signal analyzers, which are ubiquitous tools for signal processing.

## 4.1   Vibration Sensors

*Piezoelectric accelerometers* are most widely used in vibration studies. Their action is based on inertial properties of the seismic mass and electrical properties of the piezo crystal. Principal schematic of an accelerometer is shown in Fig. 4.1. The seismic mass is clamped to the piezo crystal and the base of the accelerometer. The base is attached to the object of measurement.

When the base moves along the axis of the device, the inertia of the mass generates the force that acts upon the piezoelectric crystal and causes the electric potential difference between the opposite

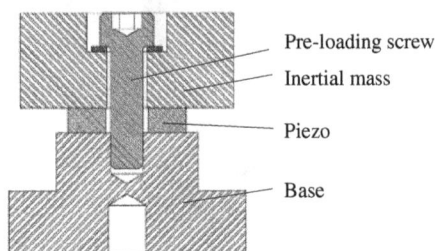

**Figure 4.1.** Principal design of a piezoelectric accelerometer (compression mode).

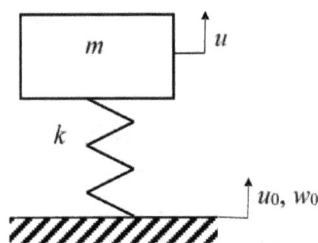

**Figure 4.2.** Single-degree-of freedom model of a piezoelectric accelerometer.

faces of the crystal that are equipped with electrical contacts. The resulting voltage is amplified and supplied to the measurement apparatus. Figure 4.1 shows a piezo crystal working in compression; other designs use crystals in shear or bending deformation modes. Some of the state-of-the-art accelerometers have built-in pre-amp and signal conditioning electronics; others require separate power supply and conditioning units.

The simplest dynamic model of an accelerometer is the familiar single-degree-of-freedom model of Chapter 3 as shown in Fig. 4.2.

The law of piezoelectricity relates the charge on the piezo crystal to the force:

$$q = d_{33}F. \tag{4.1}$$

Here $d_{33}$ is the piezo coefficient for axial deformation. On the other hand, the force is proportional to the relative displacement of the mass:

$$F = k(u - u_0),$$

and the equation of motion in the frequency domain is

$$-m\omega^2 u + k(u - u_0) = 0.$$

Recalling that the natural frequency is $\omega_0 = \sqrt{k/m}$, and acceleration of the base is $w_0 = -\omega^2 u_0$, one obtains the following relation between the force and the acceleration of the base:

$$F = \frac{m w_0}{1 - \frac{\omega^2}{\omega_0^2}}. \tag{4.2}$$

At frequencies much below the resonance, $\omega \ll \omega_0$, according to Eqs. (4.1) and (4.2), the potential difference on the faces of the piezoelectric crystal, $e$, is given by

$$e = \frac{d_{33}}{c_E} m w_0,$$

where $c_E$ is the capacitance of the crystal. Therefore, at frequencies significantly below the resonance, the device produces voltage proportional to the acceleration of the base. If an accelerometer uses, instead of axial, shear or transversal deformation, its sensitivity is based on different piezoelectric coefficients. At higher frequencies, the sensitivity of the accelerometer deviates from nominal as shown in Fig. 4.3. The chart shows the frequency ranges, traditionally noted in specifications for accelerometers, where the deviation equals 10% and 3 dB. The 10% range covers frequencies up to 30%, and the 3 dB range — up to 54% of the mounted resonance frequency. For state-of-the-art accelerometers, internal resonance frequency can be of the order of thousands Hz; on the other hand, high-sensitivity low-noise low-frequency accelerometers, employing large inertial masses, can have resonance frequencies of several hundred Hz. It is worth noting that the stiffness $k$ defining the resulting resonance includes not only the internal stiffness of the accelerometer but also the stiffness of the attachment of the device to the object under measurement. That is why methods of installation are very important. The ASA/ANSI standard [1] lists five methods of accelerometer installation that are illustrated in Fig. 4.4.

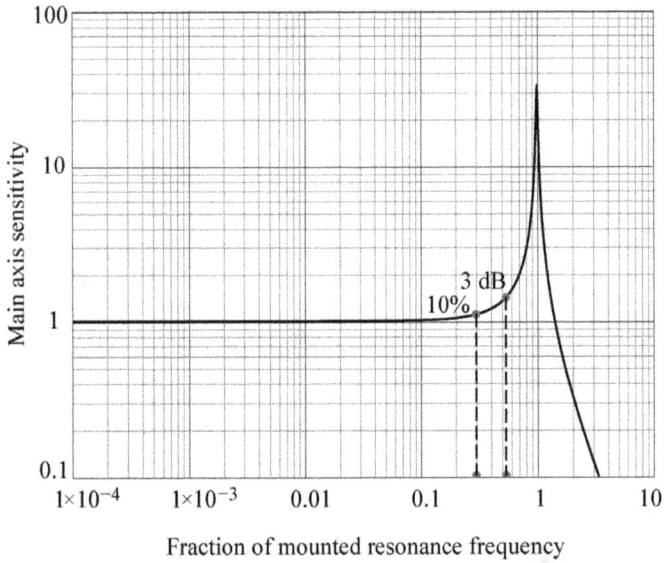

**Figure 4.3.** Accelerometer charge or voltage sensitivity as function of frequency. Dashed lines show the 3 dB range and the 10% range.

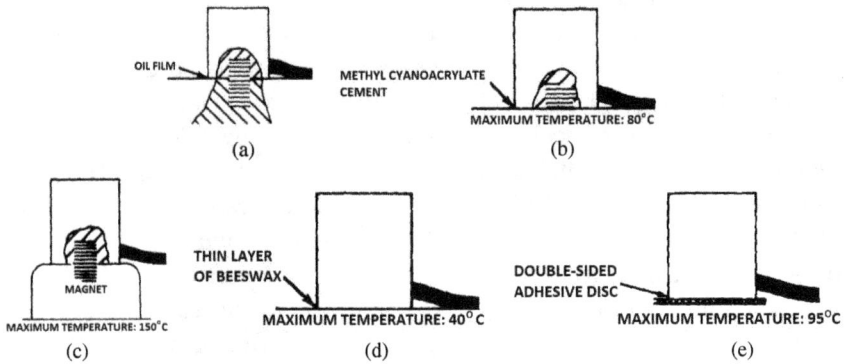

**Figure 4.4.** Methods of accelerometer installation: (a) stud; (b) cement; (c) magnet (d) wax and (e) adhesive film. Reproduced with permission from [1].

The stud or a thread adapter is the most reliable method providing a stiff connection. Most standard accelerometers have a threaded installation socket and come with a stud. A thin film of oil is sometimes added to improve contact with a rough surface. An evident limitation is that the part under measurement must have a corresponding threaded hole. Connection with wax is very popular for measurements at gravity-added attachments and moderate temperatures. It is easily installed and removed; a thin layer of wax fills the voids of a rough surface. The resulting stiffness may be less than with other methods, but usually sufficient for measurements in the range of hundreds of Hz. Methyl cyanoacrylate glues are very stiff, but, on the downside, removing the accelerometer and cleaning the surface may be not easy. Magnetic bases can be used if the surface is magnetic. It is recommended to use magnetic bases with flat surfaces and large support areas recommended by the accelerometer manufacturers; commercial off-the-shelf magnets may not provide sufficient contact stiffness. The downside of the magnetic mounting is that it introduces a foreign body between the accelerometer and the object of measurement. Double-sided adhesive tapes should be used with caution since they can be soft and heavily damped.

Another important thing to pay attention to when you install an accelerometer, as well as any vibration sensor, is cable management. To avoid the triboelectric effect (electric charge due to friction in the cable) you must exclude stressed, vibrating, or dangling cables. The cable should be taped to the measurement surface, and service loops should be arranged for accelerometers with axial connectors (see Fig. 4.5).

High-precision measurements that are often necessary for optomechanical applications put stringent requirements on the accelerometer specification. High-precision accelerometers for laboratory use produce 10 V per $g$, which can be further amplified by custom power units [2]. Power units provide the bias DC voltage necessary for operating a piezo accelerometer. Battery-powered amplifiers are often used to reduce electrical noise.

State-of-the-art precision accelerometers have equivalent electrical noise specifications both in broadband format and in terms of PSD (see Section 2.5 of Chapter 2), the latter being frequency dependent. For example, low-noise seismic accelerometer 731 A from Wilcoxon Research has specified broadband equivalent electrical

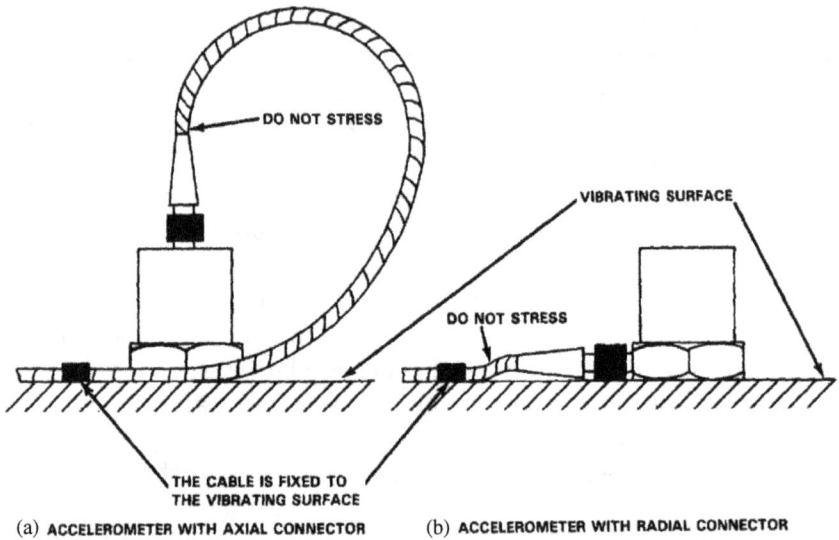

(a) ACCELEROMETER WITH AXIAL CONNECTOR        (b) ACCELEROMETER WITH RADIAL CONNECTOR

**Figure 4.5.** Accelerometer cable management. Reproduced with permission from [1].

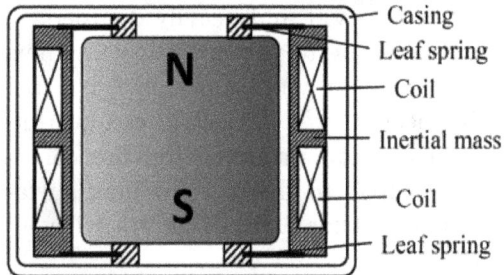

**Figure 4.6.** Principal design of a geophone.

noise 0.5 $\mu$g in the 2.5 Hz to 25 kHz range, with square root PSD of 0.03 $\mu$g/$\sqrt{\text{Hz}}$ at 2 Hz, 0.01 $\mu$g/$\sqrt{\text{Hz}}$ at 10 Hz, and 0.004 $\mu$g/$\sqrt{\text{Hz}}$ at 100 Hz [3].

*Geophone* is an electromagnetic velocity sensor. Contrary to the accelerometer, it is self-powered and does not require a bias voltage. Another difference is that the sensitivity of the geophone is, as shown below, rolled off at low frequencies, not at high frequencies. Typical design of the geophone is illustrated in Fig. 4.6, and its electromechanical model is shown in Fig. 4.7.

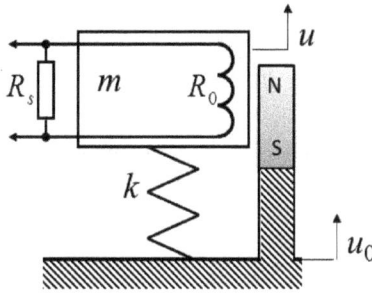

**Figure 4.7.** Electromechanical model of a geophone.

Inertial mass, which bears a coil, is attached to the casing by leaf springs. When the casing moves, the coil moves in the magnetic field created by a permanent magnet, and the voltage is generated proportional to the velocity of this motion. To derive the frequency dependence of the geophone sensitivity, note first the relation between the voltage (called "electromotive force" in this case) and the relative motion:

$$E = -Bl(\dot{u} - \dot{u}_0), \tag{4.3}$$

where $B$ is the magnetic flux density, and $l$ is the length of the electrical conductor. The product $Bl$ is called the *head constant*; it defines the efficiency of the device.

The equation of motion of the inertial mass is

$$m\ddot{u} + ku = F_e + ku_0,$$

where $k$ is the stiffness of the spring, and $F_e$ is the magnetic force that is, in turn, proportional to the current:

$$F_e = Bl \cdot I = Bl\frac{E}{R}. \tag{4.4}$$

Here $R = R_0 + R_s$, $R_0$ being the resistance of the coil, $R_s$ the resistance of the shunt (see Fig. 4.7). The inductance of a geophone coil is usually small and can be neglected in this analysis. It is important to note that $Bl$ in Eq. (4.4) is the same as in Eq. (4.3). Finally, since the output voltage is

$$U_{\text{out}} = \frac{R_s}{R} E,$$

one finds the following relationship between the output voltage and the velocity of the casing, $v_0$:

$$U_{\text{out}} = \frac{R_s}{R_0 + R_s} \frac{Bl\omega^2}{\omega^2 - \omega_0^2 \left(1 + 2i\zeta\frac{\omega}{\omega_0}\right)} v_0.$$

The output voltage is proportional to velocity if $\omega \gg \omega_0$. The effective damping, $\zeta$, consists of electromagnetic damping and mechanical damping of the spring:

$$\zeta = \frac{(Bl)^2}{2\omega_0 mR} + \zeta_{\text{mech}}.$$

The shunt resistance affects the overall sensitivity only moderately, but it strongly affects the resonance amplification. This is illustrated in Fig. 4.8 that shows a family of sensitivity functions obtained with different shunt values. The chart uses parameters typical for small-size geophones employed in vibration control technology: $Bl = 100$ V/(m/s), $m = 16$ g, $f_0 = 14$ Hz, $R_0 = 4$ kOhm, $\zeta_{\text{mech}} = 0.24$. The sensitivity rolls off at low frequencies and may have a slight resonance maximum depending on the shunt resistance. It is flat at higher frequencies, usually up to about 1,000 Hz.

*Non-contact vibration sensors* include linear variable differential transformers (LVDT), various proximity sensors, and laser vibrometers. We'll consider laser Doppler vibrometers [4] in more detail since they are gaining acceptance in optomechanical vibration studies.

Laser Doppler vibrometer determines the velocity of the moving object from the Doppler frequency shift, $\Delta f_D$, that is generated when the laser beam is reflected from the object:

$$\Delta f_D = \frac{2v}{\lambda},$$

where $v$ is the velocity and $\lambda$ is the wavelength. The frequency shift is measured by an interferometer. Acoustic modulation by Bragg cell is used to allow for determination of the direction of the motion. The general schematic is presented in Fig. 4.9.

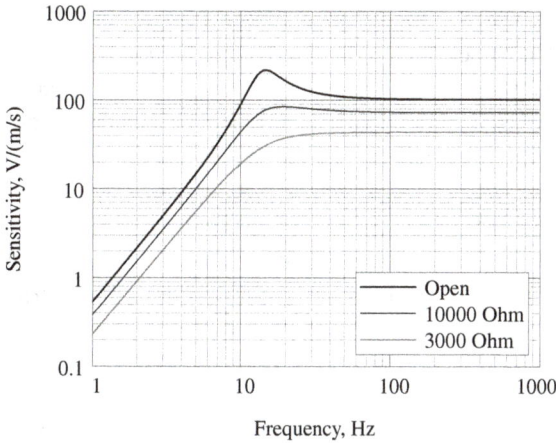

**Figure 4.8.** Frequency response of a typical small-size geophone.

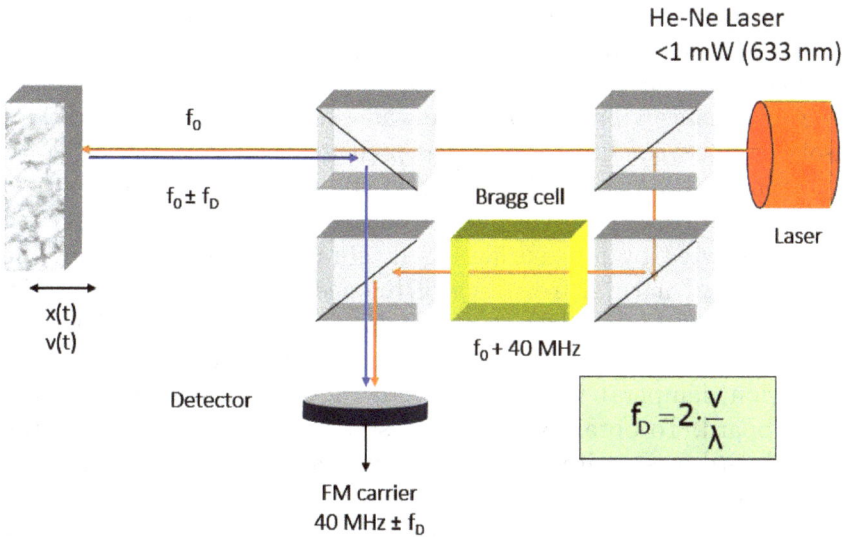

**Figure 4.9.** Principle of laser Doppler vibrometry. *Source*: Polytec, with permission.

Laser Doppler vibrometers have many advantages. They do not introduce added mass and do not require cable connections to sensors; they can be used in the scanning mode for fast collection of data from several points with automatic synchronization by a reference signal, which allows for visualization of the vibration pattern of the whole structure; they can cover wide frequency range and large amplitude range. They are not limited by the surface temperature. They can be used where other types of vibration sensors cannot, for example, for measuring vibration of objects seen through a window, remote or otherwise inaccessible objects, as well as delicate and microscopic objects.

On the other hand, application of laser Doppler vibrometry is hindered by high cost, large size and weight of the equipment, and considerable set-up time. It should be noted also that the laser vibrometer measures relative motion; therefore, steps must be taken to minimize the motion of the camera.

Application of scanning laser vibrometry is illustrated below using an example [5].

**Example 4.1. Application of scanning laser vibrometry to evaluation of the active vibration damping of a vibration-isolated platform.** The goal of the experiment was to confirm that two actively controlled electromagnetic vibration dampers can suppress resonance vibration over the whole surface of a large vibration-isolated honeycomb breadboard as predicted by the theory of active damping (see Section 9.2 of Chapter 9). A custom all-steel optical breadboard (Fig. 4.10) with dimensions 1.22 m × 1.52 m × 0.114 m was supported by four spring-based vibration isolators so that the "rigid body" natural frequency of the suspension equaled 5.25 Hz. Two SmartTable® ADD dampers (see Section 9.2 of Chapter 9 for the description of the design and principle of action of active vibration dampers) were bolted onto the two front corners of the breadboard. To obtain a consistent stationary vibration signal, the breadboard was excited by a stationary random force. A miniature electromagnetic shaker was used for this purpose; it is seen at the far-left corner of the breadboard in Fig. 4.10. The resulting vibration of the breadboard did not exceed the levels routinely observed in a laboratory environment, fitting into the VC-A vibration category (1/3-octave RMS velocity below 50 $\mu$m/s, see Section 5.2 of Chapter 5). The signal from the force sensor, inserted between the shaker and the breadboard at the driving point, served as a reference signal for the laser vibrometer data processing. An accelerometer was installed next to the shaker for auxiliary measurements.

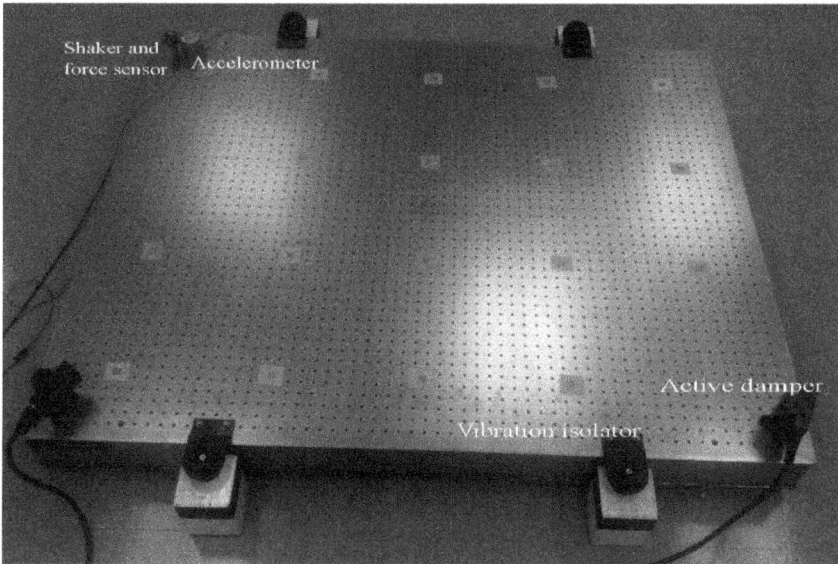

**Figure 4.10.** Optical breadboard with add-on active dampers prepared for the laser Doppler vibrometer test. Measurement points are equipped with reflecting tape.

Scanning laser vibrometer model PSV provided by Polytec was used to map the vibration of the breadboard. Figure 4.11 shows the laser scanning head installed on a gantry system and prepared for measurement. Since the breadboard was vibration-isolated, the excitation of the breadboard did not affect the camera, and it stayed essentially immovable.

Twenty measurement points were distributed uniformly over the surface and equipped with reflectors. The full scan took only a few minutes. The results of the test provided a clear picture of the damping effect over the whole surface of the breadboard. Figures 4.12(a)–4.12(d) depict the operational deflection shapes of the breadboard with active damping on and off.

Total kinetic energy is often considered the ultimate performance criterion for vibration damping systems. Scanning laser vibrometry provides an effective tool for estimating this criterion. Figure 4.13 shows the contour charts of the square root of the mean velocity squared over the surface of the breadboard. The sum of these mean

**Figure 4.11.** The laser head was attached to the gantry above the vibrating surface.

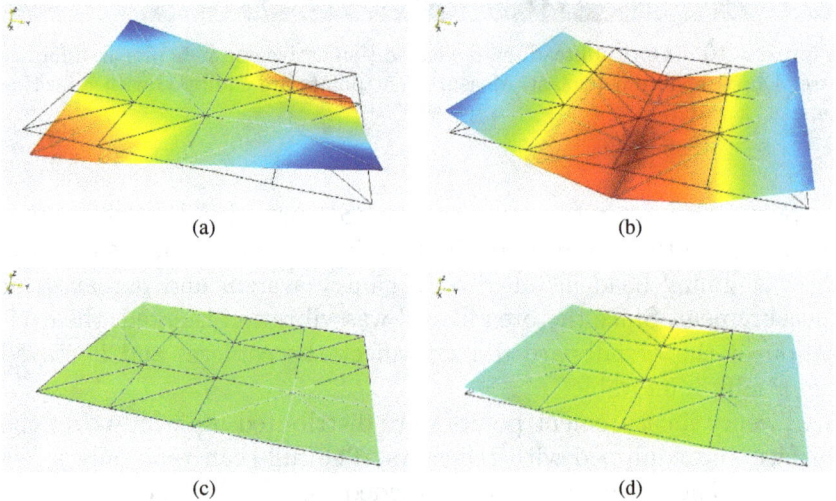

**Figure 4.12.** Operational vibration modes of the breadboard: (a) first (torsion) mode, damping off; (b) second (bending) mode, damping off; (c) first (torsion) mode, damping on; (d) second (bending) mode, damping on.

squares is proportional to the estimate of the total kinetic energy. The spectral composition of that value is plotted, in RMS units, in the graphs below. The comparison shows that the active damping practically eliminates both resonance peaks at 194 and 278 Hz, which

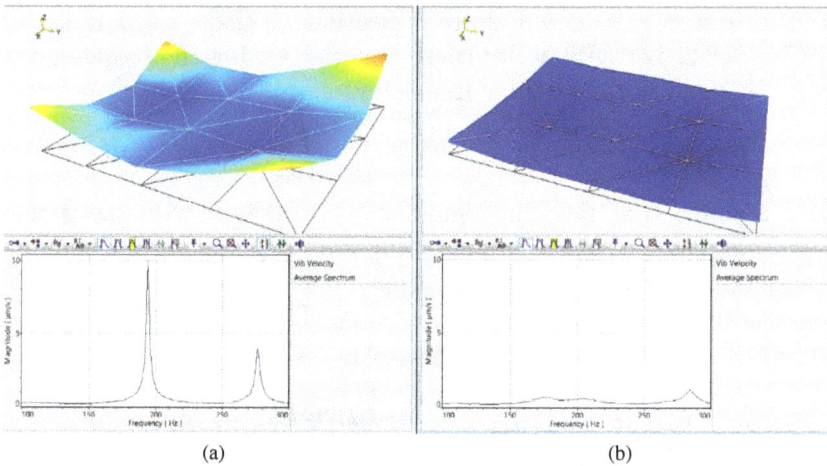

(a)                                    (b)

**Figure 4.13.** Reduction of the total kinetic energy of the breadboard by active damping. The contour charts show the RMS vibration velocity over the surface of the breadboard over the entire bandwidth of excitation. The graphs show the square root of the sum of the velocities squared over all the measurement points as function of frequency. (a) Damping off; (b) damping on.

correspond to the main twisting and bending vibration modes of the breadboard.

More details can be found in [5]. In this example, scanning laser Doppler vibrometry enabled thorough research and vivid illustration of active damping.

In conclusion of this brief survey of the most popular vibration sensors, it should be noted that no single measurement method is universal or flawless. Frequency and amplitude ranges, as well as the required sensitivity and noise floor, should be considered before choosing the most appropriate transducers.

## 4.2 Signal Analyzers: Basics of Digital Signal Processing

Signal analyzer is the central element of the measurement system. It performs four functions: data acquisition, data analysis via *discrete Fourier transform* (DFT), post-processing, and presentation.

Discrete Fourier transform is the Fourier transform performed in discrete time. The general equations for direct and inverse Fourier transform were considered above in Section 2.3 (Eq. (2.4) and forth) of Chapter 2. Let us re-write those equations using the frequency in Hz, $f$, instead of the circular frequency, $\omega$:

$$S_x(f) = \int_{-\infty}^{\infty} x(t)e^{-i2\pi ft}dt; x(t) = \int_{-\infty}^{\infty} S_x(f)e^{i2\pi ft}df.$$

DFT assumes that the time signal, $x(t)$, is periodic with the period $T = N\Delta t$. Here $\Delta t$ is the *sampling interval*, $N$ is the *block length* (an even number), and $T$ is the *record length*. The time signal is read at discrete times, $t_n = n\Delta t$, $n = 0, 1, \ldots, N$. The DFT and the inverse transform are given by the following equations [6]:

$$S_x(m\Delta f) = \Delta t \sum_{n=0}^{N-1} x(n\Delta t)e^{-i2\pi(m\Delta f)(n\Delta t)}, \quad m = 0, \ldots, \frac{N}{2};$$

$$x(n\Delta t) = \Delta f \sum_{m=0}^{N/2} S_x(m\Delta f)e^{i2\pi(m\Delta f)(n\Delta t)}, \quad n = 0, \ldots, N-1.$$

Note that, according to these equations, $N$ time points are used to calculate $L = N/2$ spectral lines. This is the result of Nyquist–Shannon–Kotelnikov theorem, which states that the *sampling frequency*,

$$f_s = \frac{1}{\Delta t}$$

should be at least two times larger than the highest frequency component present in the signal processed by DFT:

$$f_s = 2F_{\max}, F_{\max} = \frac{1}{2}f_s.$$

**Exercise 4.1.** Suppose you need to calculate, via DFT, 800 lines of vibration spectrum up to 400 Hz. What should be the minimum acquisition time?

**Solution.** Given: $F_{\max} = 400$ Hz; $L = 800$; the time period is $T = N\Delta t$. According to Nyquist–Shannon–Kotelnikov theorem, $N = 2L = 1600$, $\Delta t = 1/f_s = 1/2F_{\max} = 1.25 \cdot 10^{-3}$. Therefore, $T = 2$s. Note that, generally,

$$T = \frac{L}{F_{\max}}.$$

Signal analyzer produces a complex spectrum with the number of spectral lines equal to half of the number of real-time samples.

Note that, whereas Fourier transforms are indefinite integrals of continuous functions, discrete Fourier transforms are finite sums. DFT approximates the periodic function obtained by repeating the sampled signal. This may give rise to discontinuities at the edges of the interval, and, as a result, to the Gibbs effect (Fig. 2.4 of Chapter 2). Signal analyzers avoid the Gibbs effect by *windowing*, which means multiplication by a window function that equals one or is very close to one on most of the time interval but rolls fast to zero at both ends. A small distortion of the signal introduced this way is considered less important than possible consequences of the Gibbs effect. Hanning and flattop windows are among the most widely used [6].

Another effect of digital signal processing is *aliasing*, that is, a phenomenon of high-frequency signal components showing as low-frequency components due to sampling. It is illustrated in Fig. 4.14. To avoid it, signal analyzers employ anti-aliasing filters, which are analog low-pass filters suppressing the signals above $F_{\max}$.

Signal analyzers usually offer several *weighting* filters (A, B, and C) that are mostly used for acoustical research and are intended for bringing the signal closer to human perception. These filters are normally turned off when monitoring vibration signals.

Spectral parameters of the signal are determined by averaging over several sample functions. Special techniques of averaging are used during two-channel measurements that are performed to derive the transfer functions between the two signals (see Fig. 4.15).

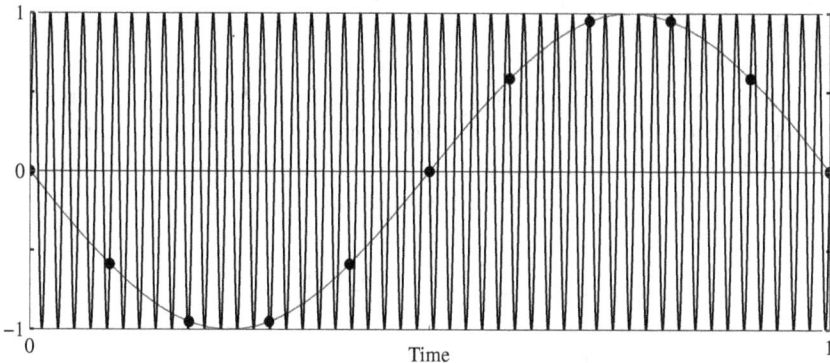

**Figure 4.14.** Aliasing: a high-frequency signal sampled at a low rate appears like a low-frequency signal.

$$F \longrightarrow \boxed{H(\omega)} \longrightarrow X$$

Channel 1         Channel 2

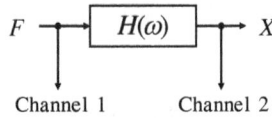

**Figure 4.15.**   Two-channel measurement.

These measurements are very important in vibration control; in particular, they are used to characterize the transmissibility of a vibration isolation system, to estimate the dynamic stiffness of a structure, or to troubleshoot a complex system by analyzing transmission of vibration from one part of the system to another (see Section 8.6 of Chapter 8).

The *transfer function* is the ratio of the output signal to the input signal in the frequency domain:

$$H(\omega) = \frac{X(\omega)}{F(\omega)}. \tag{4.5}$$

Since noise is present in all measurements, averaging is used to filter it out. Direct averaging of the ratio (4.5) is not effective for a number of mathematical reasons, so the signal analyzer calculates, by means of averaging, a cross-spectrum of input and output signals and divides it by the auto spectrum of the input signal:

$$H(\omega) = \frac{\sum F^* X}{\sum F^* F} = \frac{G_{FX}(\omega)}{G_{FF}(\omega)}.$$

This helps to filter out the uncorrelated noise in the output. In most cases there is more noise in the output than in the input. If there is more noise in the input, the procedure can be changed in evident ways. The signal analyzer calculates the best linear approximation of the relationship between the two signals. Along with the averaged transfer function, the so-called *coherence function* can be calculated that is defined by the following equation:

$$C(\omega) = \sqrt{\frac{|G_{FX}(\omega)|^2}{G_{FF}(\omega) G_{XX}(\omega)}}.$$

The coherence equals 1 if the signals are directly proportional, and it equals zero if they are completely uncorrelated. So, the coherence is an indication of the signal-to-noise ratio. The coherence will also deviate from 1 if the relation between the two signals is nonlinear. If the measurement is performed under user-generated excitation, it is recommended to check the linearity of the system by measuring the same transfer function at different input levels. It is often said that the coherence function is the measure of the quality of the data. This is not always literally true: sometimes low coherence at certain frequencies is just a consequence of the low level of signal on one or both channels at those frequencies, for example, due to the antiresonance phenomenon (see Section 3.2 of Chapter 3); in such cases, low coherence is physically substantiated and unavoidable in this kind of measurement.

In summary, here are a few practical rules to follow in vibration measurements:

- Make sure the *anti-alias* filter is on.
- Make sure the A, B, C *weighting* filters are off.
- Check the *frequency range* of analysis vs. that of analog equipment.
- Check the *noise floor* of analog equipment vs. expected levels.
- Good measurement practice: have an oscilloscope in the loop; if it is not available, display the *time domain* signals.
- Check *linearity* if possible.
- Always check *coherence*.

Measurements standards [1,7] should be followed in setting up vibration measurements.

# References

1. *Guide to the Mechanical Mounting of Accelerometers. American National Standard*, ANSI S2.61-1989 (ASA 61-1989). Reaffirmed 2001, 2005.
2. https://wilcoxon.com/wp-content/uploads/2016/07/90225-731-P31-manual.pdf. Accessed on March 1, 2021.
3. https://wilcoxon.com/wp-content/uploads/2016/08/731A-spec-98078C6.pdf. Accessed on March 1, 2021.
4. https://www.polytec.com/us/vibrometry/technology/laser-doppler-vibrometry. Accessed on October 9, 2021.
5. V.M. Ryaboy and J. Eichenberger, Keeping optical tables steady: Experimental results from a scanning laser Doppler vibrometer

elucidate the effects of active vibration damping, *Photonics Spectra*, October 2014, pp. 47–50.

6. A.J. Curtis and S.D. Lust, Concepts in vibration data analysis, C.M. Harris (ed.), in *Shock and Vibration Handbook*, 4th edn., McGraw-Hill, New York, 1995 and subsequent editions, pp. 22.1–22.34.

7. *Methods for Experimental Determination of Mechanical Mobility. Part 1: Basic Definition and Transducers. American National Standard*, ANSI S2.31-1979 (ASA 31-1979). *Part 2: Measurement Using Single-Point Translational Excitation. American National Standard*, ANSI S2.32-1982 (ASA 32-1982).

Chapter 5

# Vibration Sensitivity of Optical Instruments: Generic Criteria for Vibration Environments

Chapter 1 introduced several illustrations of the effects of vibration on optical performance. Now that we are familiar with vibration measurements and vibration spectra from Chapters 2 and 4, we are prepared to learn about vibration sensitivity of optical instruments on the quantitative level. We'll take an optical microscope as a primary example following the detailed research by Amick and Stead [1]. That will lead to introducing the standard vibration environments [2–6]. After that, we'll discuss some specific aspects of vibration sensitivity exhibited by interferometers.

## 5.1 Vibration Sensitivity of Optical Microscopes

State-of-the-art optical microscopes are complex structural systems supporting heavy optical subassemblies including appendages. This gives rise to resonance vibration responses to external impacts, sometimes at frequencies as low as 10–30 Hz. The functionality of the instrument depends critically on mutual positions and orientation of its parts. So, it comes as no surprise that the performance of optical microscopes is sensitive to vibration. It is important to characterize this sensitivity qualitatively, by settings limits to the acceptable vibration levels. These limits may be different for diverse types of microscopes and different applications.

This subject had been treated thoroughly in the article [1]. Two models of microscopes set up for 40×, 100×, 400×, and 1,000× magnifications, were subject to harmonic vibrations in vertical and horizontal directions with gradually increased amplitudes until an observer reported visible deterioration of the image. Due to the properties of human vision, this deterioration took form of "jitter" of the image at frequencies below 22 Hz and manifested as "blurriness" above this frequency. The authors found that the frequencies most sensitive to vibration were shared by different directions of excitation and clustered in the frequency range of main structural resonances. Moreover, the tests found that the threshold vibration levels were limited in the resonance frequency domain by the constant velocity amplitude lines; at lower frequencies they were limited by the constant acceleration lines and at higher frequencies — by the constant displacement lines. At all frequencies the threshold vibration levels were lower for higher magnifications and higher for lower magnifications. These findings are summarized in Fig. 5.1. It shows a

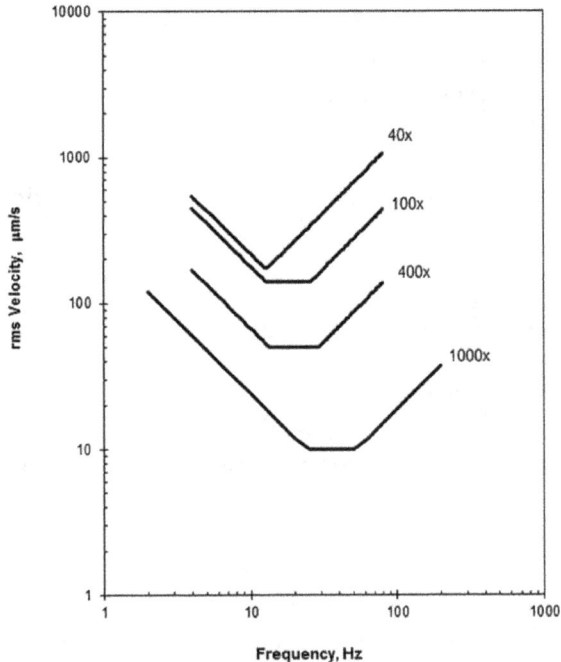

**Figure 5.1.** Vibration sensitivity thresholds for various magnifications. Reproduced with permission from [1].

series of lines marking the envelope of experimental measurement points of sensitivity thresholds of microscopes with magnifications of $40\times$, $100\times$, $400\times$, and $1{,}000\times$, as functions of frequency. These lines are formed by constant acceleration segments at lower frequencies, constant velocity segments at intermediate frequencies, and constant displacement segments at high frequencies.

Inverted microscopes with objectives below the stage pointing up have typically the same levels of vibration sensitivity as the conventional ones [1]. Articulating arms and stands reduce the resonance frequencies and increase vibration sensitivity, so these accessories are usually not compatible with high magnification. Other modifications to microscopes, required for advanced applications, can dramatically change the vibration sensitivity of the microscope-based systems. Table 5.1 summarizes the requirements to the floor or bench vibration depending on the multiplication factor and on the typical process-specific modifications. These estimates do not take into account the effect of vibration-isolating benches such as optical tables and workstations. As observed in [1], these devices can attenuate the floor vibration ten to hundred times. They will be discussed in detail in Chapters 6 and 7.

The vibration levels are interpreted in Table 5.1 through the generic criteria, called VC, recommended by the Institute of Environmental Science and Technology (IEST) [4]. These criteria were developed for general guidance in vibration sensitivity of precision instruments and processes. They are presented in detail in the next section.

**Table 5.1.** Floor or non-isolated bench vibration criteria for optical microscopes and process-specific modifications [1].

| Magnification | Floor/Bench Velocity | |
| --- | --- | --- |
| | $\mu$m/s | IEST RP12 |
| $40\times$–$100\times$ | 140 | Surgical suite |
| $400\times$ | 50 | VC-A |
| $1000\times$ | 12.5 | VC-C |
| **Process** | | |
| Digital imaging and/or fluorescence | 12.5 | VC-C |
| Microinjection, micromanipulation and/or electrophysiology | 6.3 | VC-D |
| Confocal microscopy | 6.3 | VC-D |

## 5.2    Typical Environmental Requirements (VC Criteria)

VC criteria were first developed by the consulting firm of Bolt, Beranek, and Newman [2]. They started from ISO criteria quantifying typical vibration levels in different work environments (workshop, office, residential day, operating theater), and then added lower levels, VC-A, VC-B, etc. Each VC curve is shifted down by a factor of two from the previous curve, starting from the curve marked "Operating Theater" that corresponds, according to Ungar [6], to "a most sensitive sitting or standing person's threshold of perception of vertical floor vibration."

VC criteria are stated in terms of one-third octave RMS spectral components of velocity (see Section 2.5 of Chapter 2). At lower frequencies, from 8 Hz down to 4 Hz, the requirements are relaxed according to constant acceleration lines.

Subsequently, the Institute of Environmental Science and Technology adjusted the criteria [6], following the development of new more sensitive equipment and processes, first by adding two more low levels, VC-F and VC-G, second, by prolonging high sensitive criteria, starting from VC-C, down to 1 Hz and straightening the line, thereby abandoning the criteria relaxation at low frequencies. It was done because many of the new generation of sensitive equipment had incorporated pneumatic vibration isolation of their own, which increased sensitivity at low frequencies. This is important to remember because if you deal with an instrument that does not have this built-in isolation, the low frequency requirements, which are usually hard to meet, should be relaxed. Placing the equipment with built-in pneumatic isolation on another stage of soft pneumatic isolation may lead to a two-stage system deviating from the optimal criterion of Section 3.3 of Chapter 3.

VC levels are shown graphically in Fig. 5.2. Table 5.2 contains a general description of applications suitable for each vibration level.

Many optomechanical assemblies require a specialized approach to vibration control, not limited to stating the appropriate VC level. In particular, close attention should be paid to on-board sources of vibration that may be an essential part of the application, such as fans and scanners. The moving parts should be placed on separate shelves or benches; if they need to stay close to sensitive optics, local isolation should be applied. An example of a thorough combination

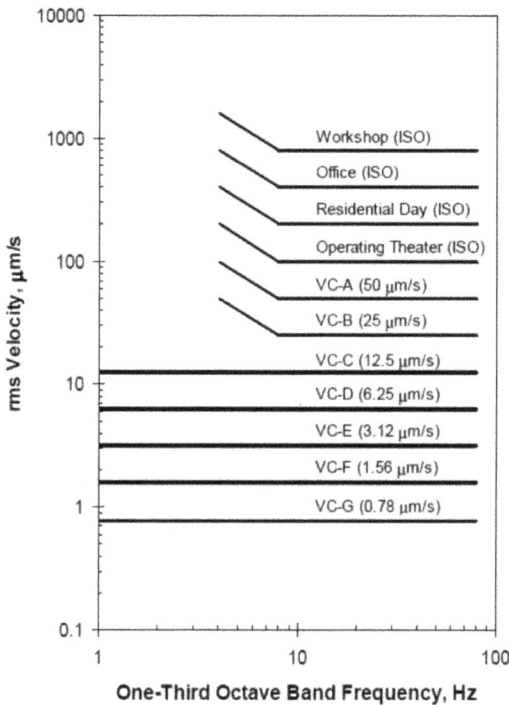

**Figure 5.2.** VC curves for vibration-sensitive equipment and ISO guidelines for buildings. Reproduced with permission from [5].

of external and on-board isolation can be found in the paper [7]. Specific challenges of measuring and reducing vibration sensitivity of optical fiber-based systems are described in [8].

Real vibration environments seldom follow the VC criteria. It should be noted that VC criteria are not intended to characterize typical vibration environments. Rather, they characterize typical *requirements* to vibration environments that are deemed appropriate to certain vibration-sensitive tasks. The VC criteria are intended for use in the frequency range up to 80 Hz, covering the typical range of the lowest resonance frequencies of optical instruments where they are most sensitive to vibration.

Attempts were made to estimate realistic vibration levels encountered in practice at the frequencies above 80 Hz. Newport Corporation (now a subsidiary of MKS) recommends the following typical levels in terms of power spectral density of floor acceleration [9]:

**Table 5.2.** Application and interpretation of the VC curves [5].

| Criterion Curve | Amplitude ($\mu$m/s) | Detail Size ($\mu$m) | Description of Use |
|---|---|---|---|
| Workshop (ISO) | 800 | N/A | Distinctly perceptible vibration. Appropriate to workshops and non-sensitive areas. |
| Office (ISO) | 400 | N/A | Perceptible vibration. Appropriate to offices and non-sensitive areas. |
| Residential day (ISO) | 200 | 75 | Barely perceptible vibration. Appropriate to sleep areas in most instances. Usually adequate for computer equipment, hospital recovery rooms, semiconductor probe test equipment, and microscopes less than 40×. |
| Operating theater (ISO) | 100 | 25 | Vibration not perceptible. Suitable in most instances for surgical suites, microscopes to 100×, and other equipment of low sensitivity. |
| VC-A | 50 | 8 | Adequate in most instances for optical microscopes to 400×, microbalances, optical balances, proximity and projection aligners, etc. |
| VC-B | 25 | 3 | Appropriate for inspection and lithography equipment (including steppers) to 3 $\mu$m line width. |
| VC-C | 12.5 | 1–3 | Appropriate standard for optical microscopes to 1000×, lithography and inspection equipment (including moderately sensitive electron microscopes) to 1 $\mu$m detail size. TFT-LCD (thin-film-transistor liquid-crystal display) stepper-scanner processes. |
| VC-D | 6.25 | 0.1–0.3 | Suitable in most instances for demanding equipment, including many electron microscopes (SEMs and TEMs) and E-beam systems. |
| VC-E | 3.12 | < 0.1 | A challenging criterion to achieve. Assumed to be adequate for the most demanding and sensitive systems including long-path, laser-based, small target systems, E-beam lithography systems working at nanometer scales, and other systems requiring extraordinary dynamic stability. |

(*Continued*)

**Table 5.2.** (*Continued*)

| Criterion Curve | Amplitude (μm/s) | Detail Size (μm) | Description of Use |
|---|---|---|---|
| VC-F | 1.56 | N/A | Appropriate for extremely quiet research spaces; generally difficult to achieve in most instances, especially cleanrooms. Not recommended for use as a design criterion, only for evaluation. |
| VC-G | 0.78 | N/A | Appropriate for extremely quiet research spaces; generally difficult to achieve in most instances, especially cleanrooms. Not recommended for use as a design criterion, only for evaluation. |

*Notes:*
[1] As measured in one-third octave bands over the frequency range 8–80 Hz (VC-A and VC-B) or 1–80 Hz (VC-C through VC-G).
[2] The detail size refers to line width in microelectronic fabrication, the particle (cell) size in medical and pharmaceutical research, etc. It is not relevant to imaging associated with probe technologies, AFMs, and nanotechnology.
*The information given in this table is for guidance only. In most instances, it is recommended that the advice of someone knowledgeable about applications and vibration requirements of the equipment and processes be sought.*

**Table 5.3.** Typical acceleration levels at high frequencies.

| Environment | Floor Acceleration PSD |
|---|---|
| Relatively quiet laboratory | $10^{-10} \ g^2/\text{Hz}$ |
| Laboratory next to street | $10^{-9} \ g^2/\text{Hz}$ |
| Light manufacturing | $10^{-8} \ g^2/\text{Hz}$ |

It should be noted that these levels refer not to the wide-band background vibration but rather to the spectral peaks related to vibration signatures of vibro-active machinery or structural resonances of the building. These peak levels are important because of the potentially negative effect they can cause if found close to the areas of increased vibration sensitivity of the optical equipment. Figure 5.3(a) shows an example of vibration levels, measured by a three-axis piezo-electric accelerometer (see Section 4.1 of Chapter 4), on a concrete floor in a laboratory close to a manufacturing site. Horizontal $x$-axis points towards the manufacturing area. Vibration on the vibration-isolated optical table at the same location (Fig. 5.3(b)) are shown

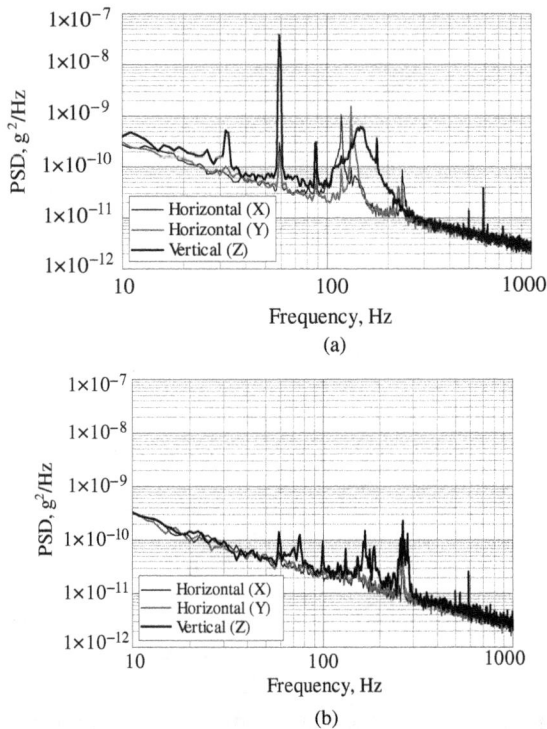

**Figure 5.3.** Vibration levels in a laboratory close to a manufacturing site: concrete floor (a) and isolated optical table (b).

for comparison. Note that the noise floor of the accelerometer is clearly seen in these graphs. Wide-band elevated vibration levels are due to people walking; sharp peaks are due to rotating electrical equipment. Isolated optical table reduces vibration significantly; however, some amplification can be noticed in local frequency bands due to increased sensitivity to acoustical noise and mechanical resonance properties of the table. Dynamics of vibration-isolated platforms is further explained in Chapters 6 and 7.

## 5.3 Vibration Sensitivity of Interferometers

The sensitivity of interferometers to vibration is well known. To quote De Groot [10], "Interferometers are almost invariably equipped with a more-or-less elaborate system to dampen or to isolate vibration.

Often the interferometer is constructed of massive machined parts and is placed on a granite slab or an air table. Auxiliary equipment, such as motors and cooling fans, is removed or is kept at a distance. No one is supposed to take a step or to breathe too heavily during the measurement process... The sensitivity of interferometers to vibration is rooted in the fundamental principles of the instrument."

All four mechanisms of vibration disturbance listed in Section 1.2 of Chapter 1 apply to interferometers. Specifically, in interferometric applications there may be two types of detrimental effects caused by mechanical vibrations: the interfering type, leading to unwanted addition to interferograms, and the modifying type, influencing the sensitivity of the interferometers [11]. As noticed in [12], mechanical vibrations have two main effects on the spectra: the addition of signals due to direct sensitivity of the detectors to vibrations (e.g., through piezoelectric effect) and the changes of the interferogram due to the motion of optical components of the interferometer.

An experiment characterizing the effect of vibration on an interferometer was conducted by Comolli and Saggin [11]. The instrument, supported by vibration isolators, was placed on a small shaker table and subjected to vibrations along three coordinate axes. Various effects of the vibrations were analyzed by comparing the IR spectra taken with and without vibration. The results under harmonic excitation showed that the sensitivity of the spectrometer to vibration is strongly dependent on frequency, increasing with frequency up to 500 Hz and decreasing at higher frequencies.

Dependence of vibration sensitivity on frequency is typical for interferometry, but the character of this dependence may be very different for different applications. For example, phase-shifting interferometers that employ mechanical displacement in the measuring process and take considerable time to build the interferogram, demonstrate complex patterns of vibration sensitivity with zones of increased and decreased sensitivity; moreover, these patterns depend on the data processing algorithm [10].

## 5.4 Vibration Control Decision Making

In conclusion to this chapter, let us summarize the main steps of the vibration control decision-making process. Whether you are designing an optomechanical device or setting up an application, the following steps should be followed as a general rule.

First, understand the vibration sensitivities of your design. Generic VC criteria are a good starting point; however, you should analyze the equipment manufacturer's data, and perform dynamic modeling and analysis as detailed as possible. Models of Chapter 3 may help in these efforts. Draw from your own and your colleague's experience and published information, and collect as much test data as you can.

Second, perform a vibration survey of the installation site or identify a typical vibration environment. Remember to use adequate measuring equipment with respect to frequency range, sensitivity, and noise specifications discussed above in Chapter 4.

Finally, decide on the necessary vibration control means that would ensure that, in the given environment, vibration impact on your application will be within the allowable limits. Vibration control methods and devices are described in the next chapters.

## References

1. H. Amick and M. Stead, Vibration sensitivity of laboratory bench microscopes, *Sound and Vibration*, pp. 10–16, February 2007.
2. E.E. Ungar, D.H. Sturz, and H. Amick, Vibration control design of high technology facilities, *Sound and Vibration*, pp. 20–27, July 1990.
3. C.G. Gordon, Generic vibration criteria for vibration-sensitive equipment, *Optomechanical Engineering and Vibration Control, Proc. SPIE Vol. 3786*, Denver, CO, 1999, pp. 22–35.
4. Institute of Environmental Sciences (IEST), Appendix C in Recommended Practice RP-012, "Considerations in Clean Room Design," *IES-RP-CC012.2*, Institute of Environmental Sciences, Rolling Meadows, IL, 2005.
5. H. Amick, M. Gendreau, T. Busch, and C. Gordon, Evolving criteria for research facilities: I – Vibration, *Proc. SPIE Conference 5933: Buildings for Nanoscale Research and Beyond.* San Diego, CA, 2005.
6. E.E. Ungar, From guess to gospel – the curious history of floor vibration criteria, *Sound and Vibration,* October 2016, p. 4.
7. A. Trache and S.-M. Lim, Integrated microscopy for real-time imaging of mechanotransduction studies in live cells. *Journal of Biomedical Optics*, vol. 14, no. 3, pp. 034024-1–034024-13.

8. A. Hati, C.W. Nelson, and D.A. Howe, Vibration sensitivity of optical components: A survey, in *2011 Joint Conference of the IEEE International Frequency Control and the European Frequency and Time Forum (FCS) Proceedings*, San Francisco, CA, USA, 2011, pp. 1–4, doi: 10.1109/FCS.2011.5977860.

9. *Newport Resource*, Catalog, Newport Corporation, 2004 and later editions.

10. P.J. De Groot, Vibration in phase-shifting interferometry, *Journal of the Optical Society of America*, vol. 12, no. 2, 1995, pp. 354–365.

11. L. Comolli and B. Saggin, Evaluation of the sensitivity to mechanical vibrations of an IR Fourier spectrometer. *Review of Scientific Instruments*, vol. 76, (2005) 123112.

12. B. Saggin, L. Comolli, and V. Formisano, Mechanical disturbances in Fourier spectrometers, *Applied Optics*, vol. 46, no. 22, (2007) pp. 5248–5256, doi.org/10.1364/AO.46.005248.

# Chapter 6

# Vibration Isolation

This chapter reviews one of the main methods of vibration control — vibration isolation — and describes the properties of several types of isolators based on springs, elastomers, and pneumatic chambers.

## 6.1 Principles of Vibration Isolation

Isolation, along with vibration damping and reduction at the source, is one of the principal methods of vibration control. Vibration isolation implies creating a barrier between the environment and the sensitive equipment that prevents vibration from reaching the equipment. An optical table (Fig. 6.1) is a prime example of vibration isolation for sensitive optical equipment. The table consists of rigid

**Figure 6.1.** An optical table on pneumatic isolators is a prime example of a vibration isolation system. Courtesy of Newport Corporation. All rights reserved.

121

tabletop and soft isolators, or legs, represented by pneumatic isolators in Fig. 6.1.

In this case we consider the floor as the source of vibration, the sensitive equipment is placed on the tabletop, and the isolators form a barrier. Sometimes the table takes the shape of an integrated system or an ergonomic workstation (Fig. 6.2), in which case isolators are integrated into the frame that supports the tabletop.

In the general case of a platform carrying sensitive equipment (Fig. 6.3), the efficiency of vibration isolators is estimated by comparing the vibration levels in isolated and non-isolated cases. Let us assume that the floor is the source of vibration, with vibration level $u_{floor}$. Isolators are introduced between the floor and the platform with the goal of reducing the vibration of the isolated equipment, $u_{isolated}$. To estimate how effective they are, we need to compare this level to the level we would have in case there were no isolators. So, we need to consider a largely hypothetical case when the platform is rigidly connected to the floor, and compare the values of $u_{rigid}$ and $u_{isolated}$. Their ratio measures the effect of vibration isolation:

$$E = \frac{u_{rigid}}{u_{isolated}}.$$

This value is called isolation efficiency, or, sometimes, insertion ratio or response ratio. In most cases — almost in all cases of precision vibration isolation — the isolators are by far the softest part of the system. Therefore, the vibration level on the floor does not depend

(a)                                    (b)

**Figure 6.2.** Vibration isolation table system (a) and workstation (b). The isolators are integrated into the support frames. Courtesy of Newport Corporation.

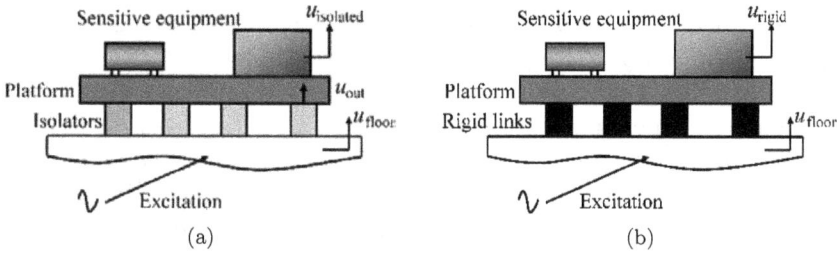

**Figure 6.3.** Isolated (a) and non-isolated (b) platforms for vibration-sensitive equipment.

on isolators, and the platform behaves, in a limited frequency range, as a rigid body. In this case, the characterization of the isolator is simplified: we can measure $E$ as the ratio of the vibration levels on the floor and above the isolators, which is the familiar transfer function (see Section 3.2 of Chapter 3) of inverse transmissibility (transmission loss) of the isolators supporting the tabletop:

$$E = \frac{u_{\text{floor}}}{u_{\text{out}}} = \frac{1}{T}.$$

The general case of non-rigid floors and isolated platforms has been treated in the literature [1]. In published specifications for isolators, foundation and isolated platform are usually assumed to be stiff enough.

The physical principles of vibration isolation, along with transmissibility functions, were discussed in Chapter 3 in the context of basic models of vibrating systems. The simplest model of the isolator is a soft viscously damped spring (Fig. 3.5 of Chapter 3). The transmissibility has been derived in Section 3.2 (see Eqs. (3.13)–(3.14) and Fig. 3.6) of Chapter 3. It indicates isolation starting from the natural frequency of the undamped mass-spring system times $\sqrt{2}$, and produces roll-off of 6 dB per octave at higher frequencies. Another, equally simple, model is a spring with a hysteretic, or frequency independent, damping. Again, the transmissibility was derived in Section 3.2 (see Eq. (3.19) and Fig. 3.11) of Chapter 3. It indicates isolation starting from the natural frequency of the undamped mass-spring system times $\sqrt{2}$, and produces roll-off of 12 dB per octave at higher frequencies. Spring-based vibration isolators largely conform to this model.

In many cases, more complicated models are used, such as a spring with complex frequency-dependent stiffness $K(\omega)$ (Fig. 3.12 of Chapter 3). It may represent a structure with multiple elastic and dissipative elements, a dual pneumatic chamber (see Section 6.4), or a visco-elastic elastomeric element. The elasticity of the isolator is characterized by the real part of the complex stiffness. It is called storage stiffness because it defines the potential energy stored in the spring. The dissipation of mechanical energy in the spring is characterized by the imaginary part of $K(\omega)$, which is called the loss stiffness, or by the ratio of the two components that is called a loss factor, see Eq. (3.21) of Chapter 3. The transmissibility of this model is given by the general Eq. (3.22) of Chapter 3.

Before we go to particular kinds of vibration isolators, let us consider briefly the test procedures for estimating their performance. The best way is to use the two-channel measurement that had been discussed in some detail in Chapter 4. One vibration sensor is installed on the floor, another one on the isolated platform, and the transfer function is measured using the averaged cross-spectra and auto-spectra, usually implemented by a dynamic signal analyzer (see Section 4.2 of Chapter 4). It should be noted that testing of an individual isolator may prove difficult since it would require a specialized test rig. So, isolators are usually tested in groups of 3 or 4 in symmetrical configurations that allow all isolators in the group to work under the same conditions.

The transmissibilities are usually measured in vertical and horizontal directions. Vertical transmissibility can be measured this way directly, provided the platform is symmetric. When measuring the transmissibility in the horizontal direction, one must be aware of the crosstalk the linear and rotational degrees of freedom. Suppose we are interested in the reaction of the platform to the horizontal movement of the floor. If the center of mass is higher than the support level, as it often happens, the horizontal reaction of supports will cause a rotational moment around the center of gravity, and the platform will rotate around the horizontal axis, engaging the stiffness of vertical supports (see Fig. 6.4). This is sometimes called a "rocking" mode. If the properties of the isolators in the vertical direction are known, the horizontal stiffnesses can be estimated from this test by curve-fitting or inverse calculation based on these properties and the known distribution of the mass in the platform.

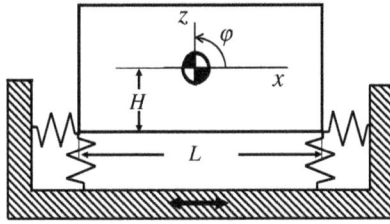

**Figure 6.4.** Horizontal and rotational vibrations are coupled.

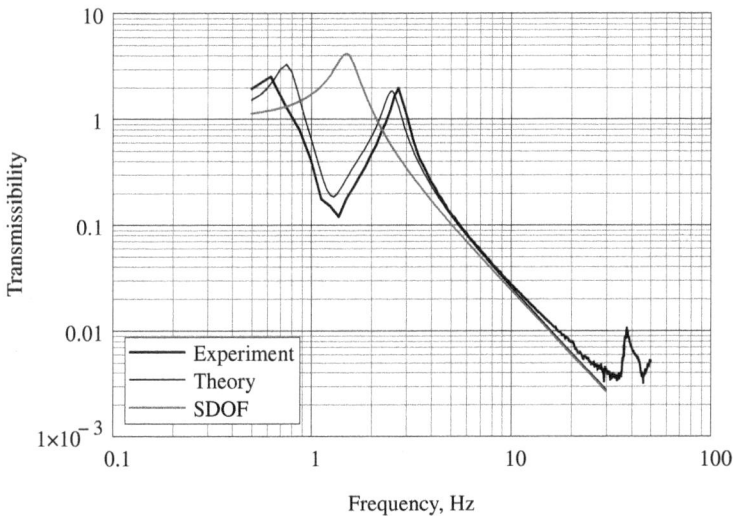

Frequency, Hz

**Figure 6.5.** Horizontal and rocking modes of a heavy platform supported by pneumatic isolators excited by horizontal vibration of the floor. Measured (thick line) and calculated based on estimated mass and moment of inertia (thin line) transmissibilities are compared to the "nominal" single-degree-of-freedom response defined by horizontal stiffness of the isolators (dim line).

Figure 6.5 shows an example of measured and calculated transfer functions of a loaded optical table, from a horizontal accelerometer installed on a shaking platform to a horizontal accelerometer installed on the table.

Section 3.4 and specifically Exercise 3.8 of Chapter 3 give additional information on vibration-isolated rigid body and rocking modes.

## 6.2 Spring Isolators: Constant Natural Frequency Property

Metal springs belong to the oldest known vibration isolation devices. The isolators based on metal springs have high load capacity, low creep, and low thermal sensitivity. They can withstand large deformations. Their dynamic stiffness is practically equal to their static stiffness (below the internal resonance frequencies). The downside is the lack of inherent damping that leads to high resonance response and long settling time after transient excitation.

The load-deflection relationship for standard constant-pitch circular helical springs (Fig. 6.6) is given by the following equation [2]:

$$P = \frac{Gd^4}{8D^3 n} u, \tag{6.1}$$

where $P$ is the axial load, $u$ is the axial displacement, $D$ is the median diameter of the coil, $d$ is the diameter of the wire, $G$ is the shear modulus of the spring material, $n$ is the number of coils. For springs with ground surfaces, $n$ is the number of active coils.

Transversal stiffness of a helical spring depends on the axial contraction. Transversal stiffness is close to axial stiffness for moderate pre-loads and height-to-diameter ratios around 1.5; transversal to axial stiffness ratio diminishes with increased height-to-diameter ratio and with increased pre-load [2].

Equation (6.1) suggests that the spring has a constant (independent of the load) stiffness that is inversely proportional to the number of active coils. Effective vibration isolation requires low stiffness, which leads to increasing the length of springs. This may cause static

**Figure 6.6.**    Constant pitch cylindrical helical spring.

instability of compression springs and internal resonance vibrations of the springs. The useful load range of helical springs is limited by settling of the coils on each other.

These shortcomings can be ameliorated by using variable stiffness springs that stiffen up with increased loads, in particular, constant natural frequency (CNF) springs that maintain the same frequency of natural vibration as the load increases. CNF springs offer wide load ranges and other advantages in vibration isolation, especially for multi-support unevenly loaded systems [2]. To derive the load-deflection characteristics necessary for the CNF property, recall that the natural frequency of the small axial vibration of the spring-supported mass $m$ is expressed by the formula for a linear oscillator (Section 3.1 of Chapter 3),

$$\omega_0 = \sqrt{\frac{k}{m}}, \quad k = m\omega_0^2. \tag{6.2}$$

The load on the spring is $P = mg$, where $g$ is the acceleration of gravity. For a nonlinear spring, the stiffness depends on the load. The natural frequency of small vibration is calculated using the tangential stiffness, $k = dP/du$. The requirement that $\omega_0$ be constant leads, therefore, to the following differential equation for the load-deflection curve:

$$\frac{dP}{du} = \frac{P}{g}\omega_0^2, \tag{6.3}$$

with the general solution in the form

$$P = P_0 e^{\frac{\omega_0^2}{g}(u-u_0)}. \tag{6.4}$$

A complete load-deflection curve of any spring must pass through the point $P = 0$, $u = 0$. Unfortunately, the solutions (6.4) do not have this property. That means that no elastic element can follow the CNF rule at all loads starting from zero. A practical design [3] should follow a linear load-deflection relationship at low loads,

$$P = \frac{P_0}{u_0}u \tag{6.5}$$

for $P < P_0$ and follow the CNF requirement (6.3) for $P \geq P_0$. The condition of smooth transition between the two load ranges (continuity of the tangential stiffness $k = dP/du$ at $u = u_0$) leads to

$$u_0 = \frac{g}{\omega_0^2}, \tag{6.6}$$

which finalizes the definition of the ideal CNF load–deflection curve: the load $P$ and the deflection $u$ of a nonlinear elastic element that supports vertical load providing constant natural frequency $\omega_0$ at $P \geq P_0$ and a linear characteristic at $P < P_0$, are related by equations (6.4) for $u \geq u_0$ and (6.5) for $u < u_0$, where $u_0$ is given by Eq. (6.6). Figures 6.7 and 6.8 illustrate the ideal CNF load–deflection relationships.

Nonlinear springs, including CNF springs, can be designed as variable pitch cylindrical helical springs (Fig. 6.9). The coils with smaller pitch settle on each other first, thereby reducing the effective number of active coils and increasing the stiffness of the spring. Another method is based on variable diameter springs (Fig. 6.10). In this case, larger diameter (softer) coils settle and become inactive first, thus increasing the stiffness. Conical constant pitch springs, Fig. 6.10(a), have stiffness increasing with load too fast to comply with the CNF law. It was shown [3] that a spring must have a logarithmic profile, Fig. 6.10(b), to follow the CNF rule exactly. CNF springs with

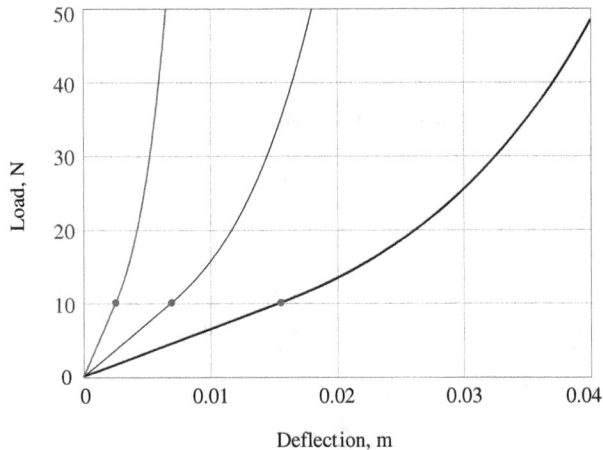

**Figure 6.7.** Ideal CNF load–deflection curves providing constant natural frequencies at loads higher than 10 N: 4 Hz (thick line), 6 Hz (thin line), 10 Hz (dim line). Dots mark the beginnings of CNF portions of the curves.

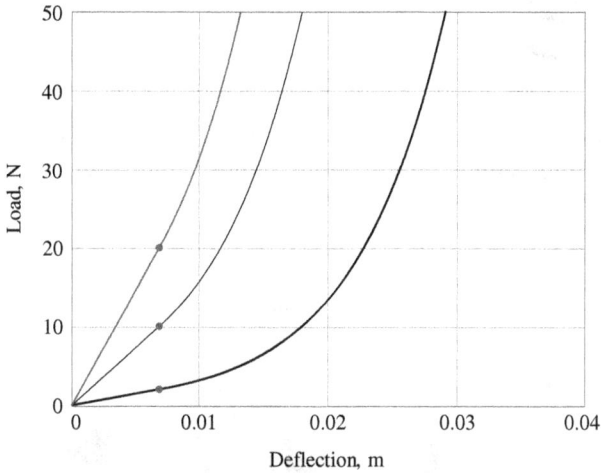

**Figure 6.8.** Ideal CNF load–deflection curves providing constant natural frequencies equal 6 Hz at loads higher than 2 N (thick line), 10 N (thin line), 20 N (dim line). Dots mark the beginnings of CNF portions of the curves.

**Figure 6.9.** Variable pitch cylindrical helical spring.

logarithmic profiles are called Iorish's isolators, by the name of the inventor [3,4]. Such springs may be difficult to manufacture; simpler profiles such as combinations of the conical and cylindrical parts, Fig. 6.10(c), can follow the CNF property closely enough for practical purposes.

Although coil springs are most widely used, other types of springs, e.g., slotted springs [2], Belleville springs [2], and wave springs [5] find their applications. Spring-based isolators are used in isolated platforms for sensitive equipment such as microscopes.

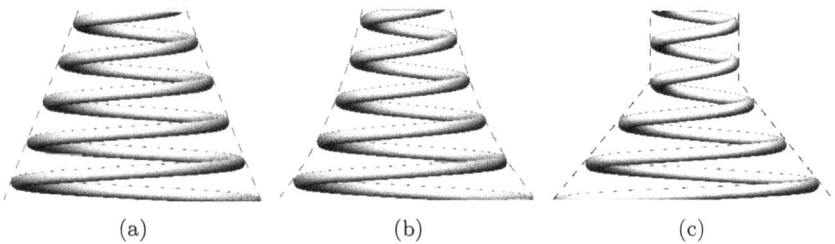

(a)                          (b)                          (c)

**Figure 6.10.**   Variable diameter coil springs: (a) conical spring; (b) Iorish's CNF spring; (c) spring combining cylindrical and conical parts.

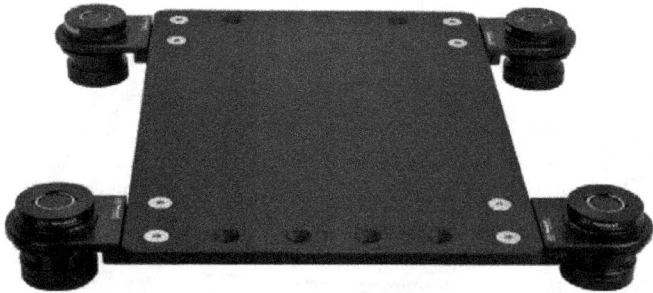

**Figure 6.11.**   A spring-based vibration-isolated platform for sensitive equipment [6]. Courtesy of Newport Corporation.

Figure 6.11 shows an example of a spring-based platform. Typical transmissibilities achievable by this system, based on variable-pitch wave springs, are shown in Fig. 6.12; they demonstrate roll-off of about 12 dB/octave compatible with frequency-independent damping (see Section 3.2 of Chapter 3). Practical realization of these designs gives results close to ideal CNF as illustrated by Fig. 6.12 (bottom row).

A special class of spring-based vibration isolators is represented by highly nonlinear so-called "negative stiffness" isolators that combine traditional springs with potentially unstable spring elements to achieve very low stiffness at a particular load. This way, excellent vibration isolation can be combined with sufficient load capacity. The principle of action of these isolators is illustrated in Fig. 6.13. An elastic beam shaped like an arch (Fig. 6.13, top left) has the ability to "snap through" as the load increases. Its load–deflection curve (Fig. 6.13, top right) has a portion with negative stiffness.

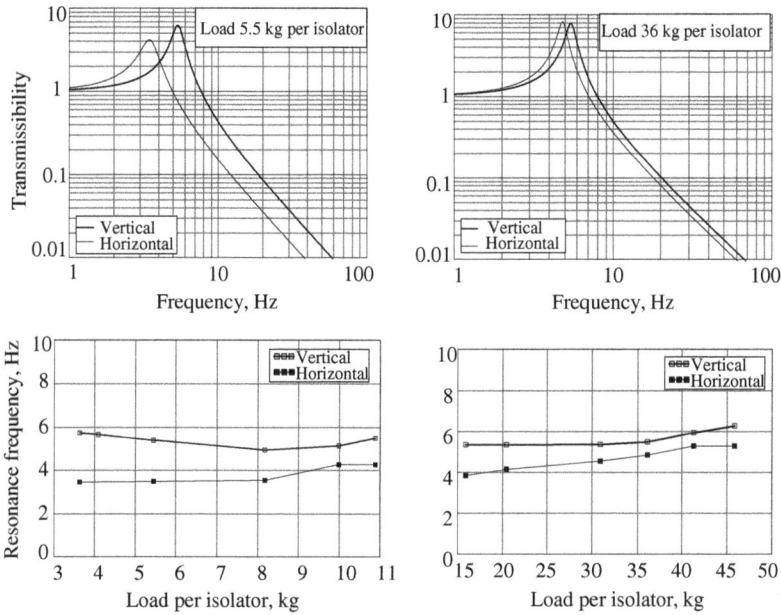

**Figure 6.12.** Transmissibilities and resonance frequencies of two models of variable pitch CNF wave spring-based vibration isolation platforms. Courtesy of Newport Corporation. All rights reserved.

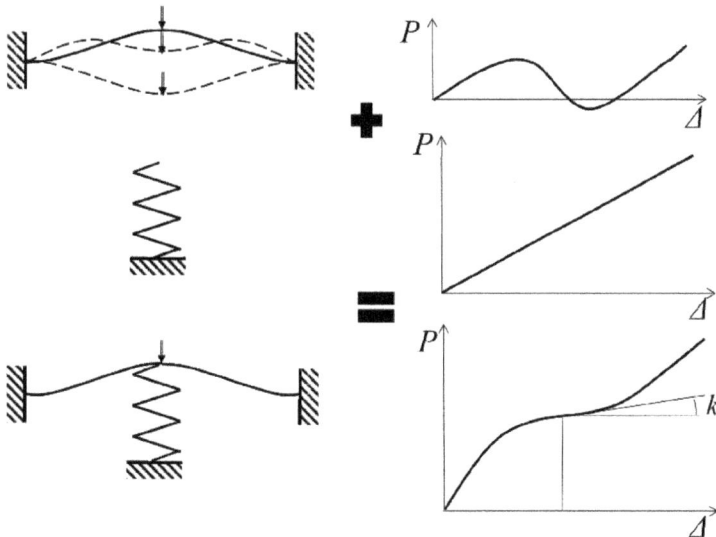

**Figure 6.13.** Principle of action of "negative stiffness" vibration isolators.

Connecting this unstable element in parallel with a traditional spring that has stiffness only slightly exceeding the absolute value of this negative stiffness (Fig. 6.13, middle row) results in nonlinear load–deflection characteristics with an inflection point marking very low stiffness $k$ in a local load range (Fig. 6.13, bottom right).

Detailed dynamic analysis and design principles of this class of vibration isolators can be found in the monograph by Alabuzhev *et al.* [7]. A practical design must include a precise adjustment mechanism to keep the low-stiffness part of the load–deflection curve close to the supported load. An isolator invented by Platus [8], with an unstable element in the form of a vertical column, found applications in precision vibration control of optomechanical systems. Vibration isolation with natural frequency as low as 0.5 Hz was demonstrated by this device.

Certain precautions should be observed with these isolators. They are strongly nonlinear, so frequencies absent in the input vibration may appear in the output (super-harmonics and sub-harmonics). Also, they are very sensitive to environmental conditions, so constant manual adjustment may be necessary to maintain the operational range.

## 6.3 Elastomeric Isolators

Elastomeric isolators for precision vibration control are based on resilient elements manufactured from highly engineered polymer materials, mostly polyurethanes or neoprene rubbers. To be useable in optomechanical applications, these materials must combine the demanding properties of high damping, low creep, and low outgassing.

The elastomeric elements of vibration isolators can take different shapes, depending on the requirements to axial and transversal stiffness. Cylindrical, conical, and tapered elements are shown in Fig. 6.14; sometimes more intricate shapes are used [2].

Mechanical properties of elastomers are strongly dependent on frequency and temperature. Typically, elastic moduli increase with frequency. The loss factor, on the contrary, reaches its maximum and stays stationary in the mid-frequency area. A typical example of

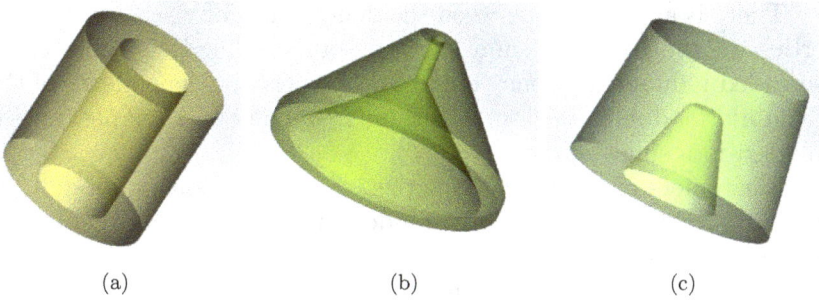

Figure 6.14. Elastomeric elements: (a) cylindrical, (b) conical, (c) tapered.

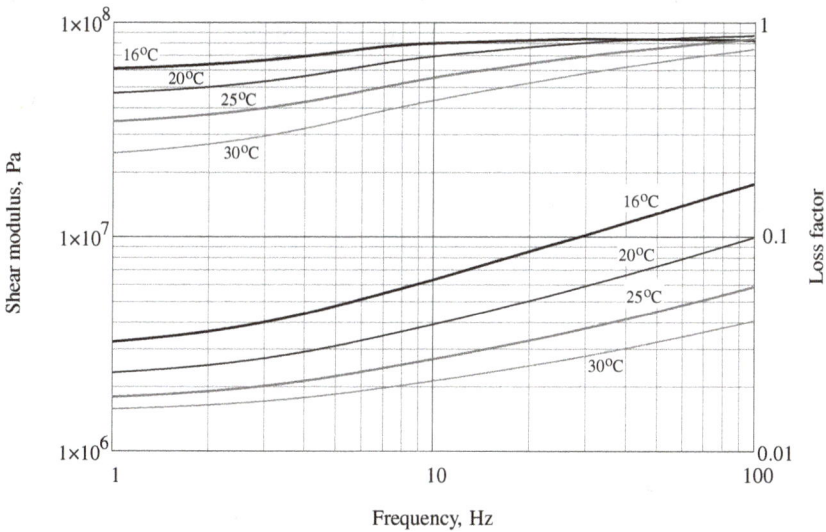

Figure 6.15. Dependence of elastic modulus (four curves at the bottom) and loss factor (four curves at the top) on frequency and temperature for a highly damped polyurethane elastomer.

frequency dependence of shear modulus and loss factor for a highly damped polyurethane elastomer is shown in Fig. 6.15. Note that in the frequency range where the damping is close to the maximum (50–100 Hz in this case), the loss factor is relatively unaffected by frequency and temperature, whereas the elastic modulus is strongly dependent on both factors.

There is an analogy between the dependency of viscoelastic properties on the frequency and the temperature [9]. It is reflected in so-called reduced frequency nomograms where the dependence of the dynamic modulus and loss factor on frequency and temperature is plotted in a single curve.

Elastomers are often characterized, instead of elastic moduli, by hardness, defined by the depth of indentation of a specially calibrated probe on the surface of the material. There is a relation between the hardness rate and the elastic modulus. This relationship is illustrated in Fig. 6.16. The Poisson ratios of elastomers are very close to 0.5.

Elastomeric isolators can be designed to have the CNF property. This can be achieved by a special geometry of the support or interface plates and/or a special form of the elastomeric element [2, 11]. Figure 6.17 shows one of these designs. As the axial load increases, the outer conical surface of the elastomer is pressed against the profiled surface of the upper support plate, thereby increasing the stiffness of the elastomer in proportion with load and achieving a CNF property. The load–deflection curve of this isolator and typical transmissibilities are shown in Fig. 6.17.

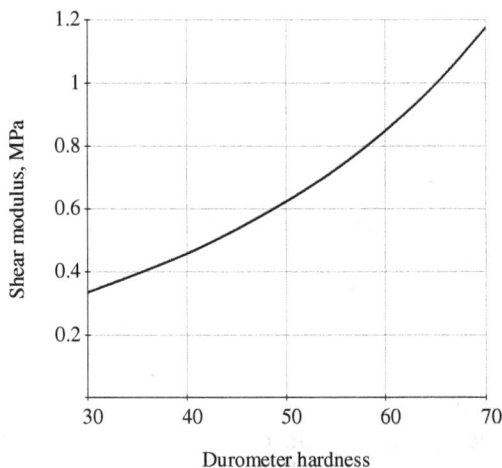

**Figure 6.16.** The relation between the hardness and the elastic modulus (after Crede [10]).

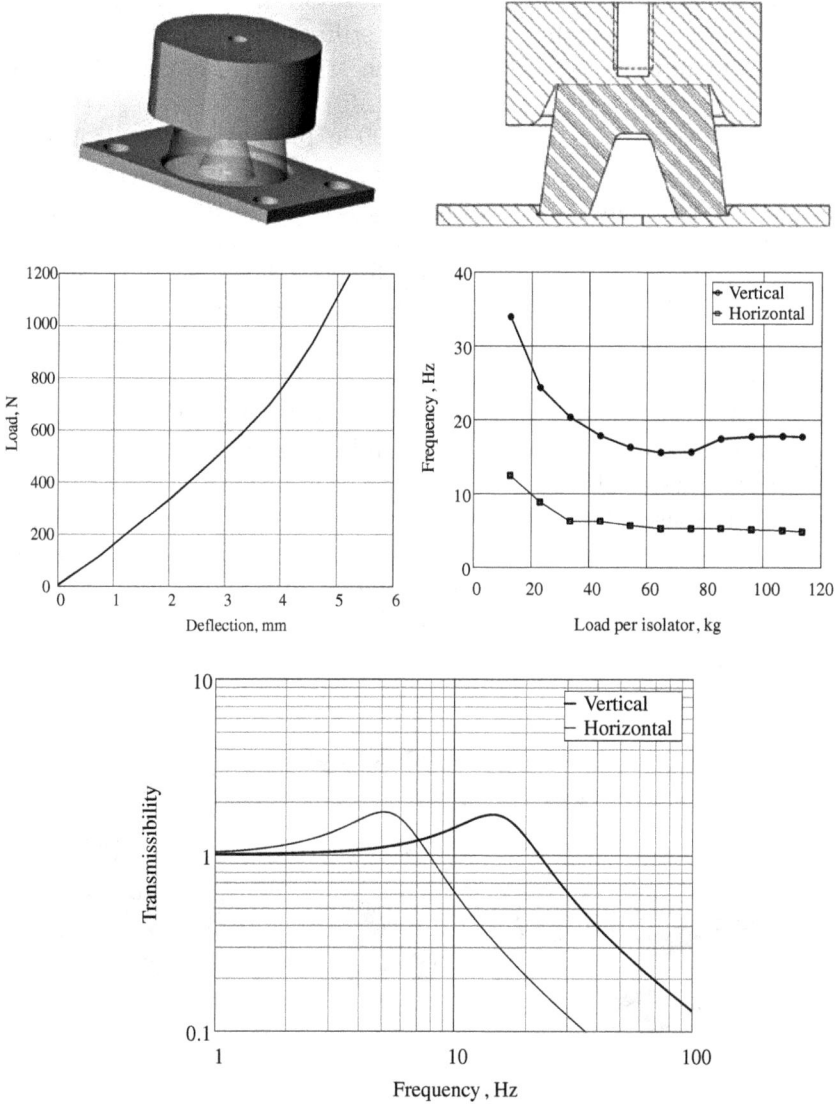

**Figure 6.17.** General appearance, load–deflection curve, natural frequencies of vertical and horizontal vibrations as functions of load, and typical transmissibility curves of the elastomeric isolator ND01-A. Courtesy of Newport Corporation. All rights reserved.

The shape of an elastomeric isolator can be chosen to obtain a desirable ratio of stiffnesses in axial and transversal directions: conical isolators usually have close stiffnesses in both directions, whereas cylindrical isolators are much stiffer in the axial (vertical) than in the transversal (horizontal) direction. This anisotropy of cylindrical elastomers can be a valuable property for motion control applications involving fast-moving air bearing stages that are sensitive to the tilt of the platform. Figure 6.18 shows an example of such a system supported by blocs of elastomeric isolators. Elastomeric isolators can present an optimal trade-off between requirements of good vibration isolation, short settling time, and small tilt, which makes them, in many cases, isolators of choice for motion applications.

Certain precautions must be taken when using elastomeric isolators. Most applications to optics and optoelectronics require low-outgassing materials. This is not a common property for elastomers. NASA database [12] can help in choosing a low-outgassing elastomer. Elastomers have high thermal expansion coefficients compared to metals, and long-term creep can happen during long-term exposure to loads; so, some means of re-leveling should be designed in the

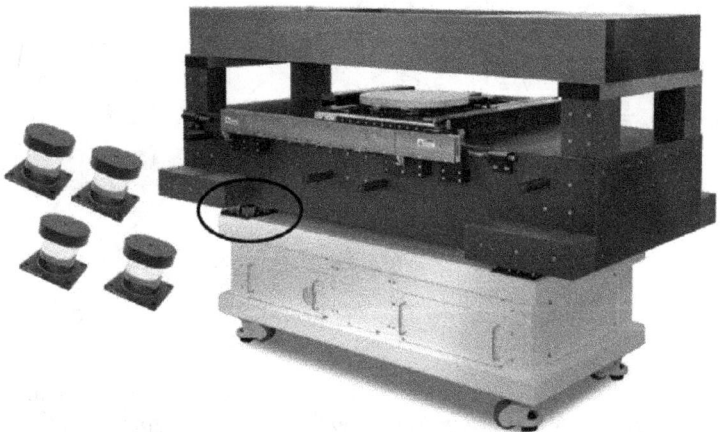

**Figure 6.18.** DynamYX® Datum TM 450 stage is supported by four sets of four highly damped cylindrical elastomeric isolators. Courtesy of Newport Corporation. All rights reserved.

applications sensitive to exact positions. If an elastomer is overloaded or accidentally damaged, a failure usually takes a long time to develop, so periodic inspection of the elastomers' condition should be included in the maintenance schedule. It is not recommended to have sources of heat near elastomeric supports; thermal shielding should be used if necessary.

## 6.4 Pneumatic Isolators

Pneumatic vibration isolator is a "workhorse" of precision vibration control. Pneumatic isolators in their modern form have been used for precision vibration control for several decades. They help meet stringent vibration control requirements in demanding application areas such as optoelectronics, life sciences, nanotechnologies, and microelectronics. Besides excellent vibration isolation characterized by natural frequencies as low as 1 Hz, they provide high load capacity that enables isolation of large setups including heavy elements like high-power lasers, atomic force microscopes, scanning tunneling microscopes, cryostats, and large optical assemblies. The basic properties of these isolators are described in this section following the paper [13].

A pneumatic chamber, also known as an acoustic cavity, is the "heart" of a pneumatic vibration isolator. The linear stiffness of the pneumatic chamber is given by the following equation:

$$K = \gamma \frac{p_0 A^2}{V}. \tag{6.7}$$

Here $p_0$ is the total pressure, $A$ is the piston area, and $V$ is the internal volume, see Fig. 6.19(a). Equation (6.7) is derived from the assumption that the compression–expansion process of air in the chamber is adiabatic. Hence, $\gamma$ is the adiabatic constant, or ratio of specific heats, equal to 1.41 for air. Both theoretical solution and experimental evidence show that, for typical sizes of air chambers of pneumatic vibration isolators, the thermal conductivity of air is not significant, and adiabatic equations may be used at all frequencies, even for slow quasi-static processes [13]. A linear model of the pneumatic chamber is valid unless the displacements are comparable to the length of the chamber [14].

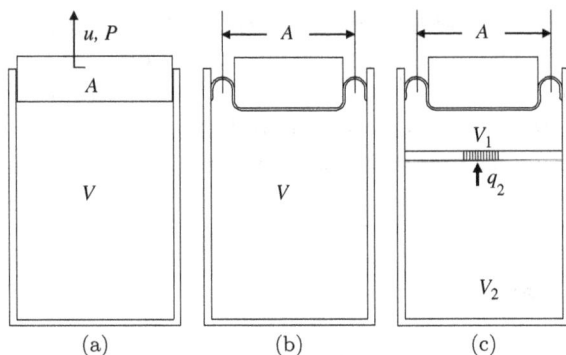

**Figure 6.19.** Schematic representations of a pneumatic vibration isolator: (a) pneumatic chamber closed by a piston, (b) pneumatic chamber with a piston sealed by a rolling diaphragm, (c) two-chamber pneumatic isolator. Implements for isolating horizontal motion are not shown in these diagrams.

Equation (6.7) illustrates one of the most valuable properties of a pneumatic isolator, namely, its ability to adjust the stiffness to the load. Since the pressure increases with the supported weight, the natural frequency tends to stay almost constant in a range of supported loads, thereby approaching the CNF property that is desirable in vibration isolation systems for a number of reasons explained in advanced texts on vibration control [2]. The CNF property was discussed in more detail in the previous sections. Since the total pressure is a sum of the gauge pressure and the atmospheric pressure, $p_0 = p_g + p_{atm}$, and the gauge pressure is defined by the load supported by the isolator, $p_g = mg/A$, the "undamped" natural frequency of the oscillations of the mass $m$ supported by the compressed air volume is, according to Eq. (3.4) of Chapter 3 and Eq. (6.7),

$$f_n = \frac{\omega_n}{2\pi} = \frac{1}{2\pi}\sqrt{\frac{K}{m}} = \frac{1}{2\pi}\sqrt{\gamma\frac{gA}{V}\left(1 + \frac{p_{atm}A}{mg}\right)}. \qquad (6.8)$$

This formula shows that, even for an ideal pneumatic chamber, the frequency has some variation with the load, especially at low loads. An example is shown in Fig. 6.20 (solid line). The value

$$\omega_* = 2\pi f_* = \sqrt{\gamma gA/V}, \qquad (6.9)$$

called the nominal circular natural frequency of the pneumatic chamber, is the liming lowest value of $\omega_n$ at high pressures.

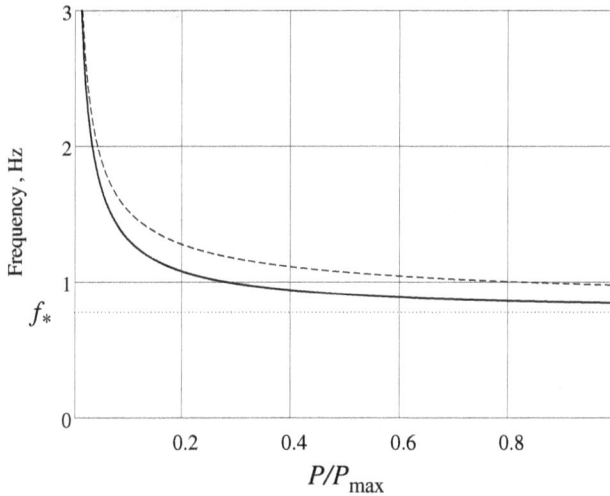

**Figure 6.20.** Natural frequency vs. load. Solid line — diaphragm excluded, dashed line — diaphragm included. $V/A = 0.59$ m, $P_{max}/A = 5.2\, p_{atm}$.

Another element that affects the stiffness of a pneumatic isolator is the diaphragm, sometimes called "rolling diaphragm", which seals the air volume and allows for free motion of the piston in the process of re-leveling. Equations (6.7)–(6.9) are still applicable with $A$ meaning the "effective" piston area which covers the space up to half-span of the inflated diaphragm, see Fig. 6.19(b). Speaking about the linear stiffness of the diaphragm, distinction should be made between two ranges of piston displacement. When the displacements are small, the diaphragm works like a linear elastic spring with very low damping. This range spans about 10 microns of relative displacement amplitude, which covers typical displacements in a laboratory environment. This value of stiffness should be used when analyzing vibration isolation performance (transmissibility) of the isolator; it is roughly proportional to the square root of the load [13]. At larger displacements, the diaphragm starts to "roll", thereby providing higher friction and lower stiffness.

Damping is necessary for isolation systems to limit the resonance amplification. To this effect, a two-chamber "self-damping" design is used in state-of-the-art pneumatic vibration isolators. The schematic is shown in Fig. 6.19(c). The air volume is divided into two chambers: the upper one, $V_1$, sometimes called a compliance chamber, and the lower one, $V_2$, called a damping chamber or a surge chamber. The two chambers are connected by a flow resistance orifice that is designed to provide a laminar viscous flow of gas between the chambers.

If the pressure in the upper chamber deviates from the equilibrium value $p_0$ by a small quantity $p_1$, and the pressure in the lower chamber deviates from $p_0$ by a small quantity $p_2$, then the volume flow of gas from $V_2$ to $V_1$, $q_2$, is described, in the linear approximation, by the equation

$$p_1 - p_2 = -c_{12}\dot{q}_2. \tag{6.10}$$

Here $c_{12}$ is the viscosity coefficient depending on the viscosity of the gas and the geometry of the orifice. To create a linear analytical model of the two-chamber pneumatic isolator, complement (6.10) with the linear adiabatic flow equations for both chambers:

$$p_1 = -\frac{\gamma p_0}{V_1}(q_1 - q_2), \quad p_2 = -\frac{\gamma p_0}{V_2}q_2. \tag{6.11}$$

Change of volume of the upper chamber, $q_1$, is caused by the small motion of the piston, $u_1$: $q_1 = Au_1$. Variation of the force acting on the isolator results from the change of pressure and the additional force exerted by the diaphragm: $P = -Ap_1 + K_d u_1$, where $K_d$ is the linear stiffness of the diaphragm (see Fig. 6.19 for the sign convention). Introducing equivalent displacement for the second chamber, $u_2 = q_2/A$, one obtains from Eqs. (6.10) and (6.11) the following system of equations for the mechanical model of a two-chamber isolator:

$$\begin{aligned} P &= K_1(u_1 - u_2) + K_d u_1, \\ K_1(u_1 - u_2) - K_2 u_2 &= C_{12}\dot{u}_2, \end{aligned} \tag{6.12}$$

where $K_1 = \gamma p_0 A^2/V_1$, $K_2 = \gamma p_0 A^2/V_2$, $C_{12} = c_{12}A^2$. This corresponds to the schematics shown in Fig. 6.21. Here $K_1$ represents the stiffness of the upper chamber, $K_2$ represents the stiffness of the lower chamber, and $C_{12}$ represents the viscosity of the gas moving through the orifice.

Equation (6.12) and Fig. 6.21 lead to the formula for the dynamic stiffness of the isolator, that is, the ratio of the amplitude of harmonic force, $P$, acting on the isolator, to the amplitude of displacement, $u_1$,

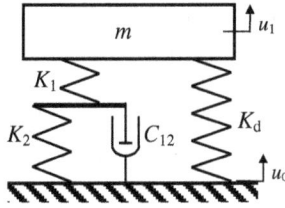

**Figure 6.21.** Mechanical model of a two-chamber pneumatic isolator.

of the isolator, as function of frequency:

$$K(\omega) = \frac{1}{\frac{1}{K_1} + \frac{1}{K_2 + i\omega C_{12}}} + K_d. \qquad (6.13)$$

This is a complex stiffness that combines elastic and dissipative components. Dynamic stiffness of the two-chamber pneumatic isolator $K(\omega)$ (6.13) has two limiting values: at low frequencies ($\omega \to 0$) the flow resistance is low, and $K(\omega)$ assumes the lowest value, $K_L$, which corresponds to both chambers acting as one:

$$K_L = \gamma p_0 A^2 / (V_1 + V_2) + K_d;$$

at high frequencies, the flow resistance grows, and $K(\omega)$ approaches the highest limiting value, $K_H$, which corresponds to the top chamber acting alone:

$$K_H = \gamma p_0 A^2 / V_1 + K_d.$$

Damping can be normalized to the time constant, $\tau$, of the pressure-equalizing between the chambers, as follows:

$$\tau = \frac{C_{12}}{K_1 + K_2}. \qquad (6.14)$$

Using these values, the dynamic stiffness can be written in the following compact form:

$$K(\omega) = \frac{K_L + K_H i\omega\tau}{1 + i\omega\tau}. \qquad (6.15)$$

The performance of the isolator is characterized by its transmissibility as explained in Section 6.1. Suppose the isolator supports an object modeled by a concentrated mass $m$ that must be protected from the vibration of the foundation. The equation of motion can be written in the frequency domain as follows:

$$[-m\omega^2 + K(\omega)]u_1 = K(\omega)u_0. \qquad (6.16)$$

Vibration isolation is characterized by the transmissibility, $T(\omega) = u_1/u_0$. From (6.16),

$$T(\omega) = \frac{K(\omega)}{-m\omega^2 + K(\omega)}. \qquad (6.17)$$

The stiffness of the diaphragm, $K_d$, is usually much lower than the stiffness of the pneumatic chamber. For practical analysis, one can assume, for sake of simplicity, that the damping provided by the diaphragm is negligible, so that $K_d$, and therefore $K_L$ and $K_H$, are real numbers. Substituting expression (6.15) into (6.17) gives the following equation for the transmissibility:

$$T(\omega) = \frac{1}{1 - \frac{\omega^2}{\omega_L^2}\frac{1+i\omega\tau}{1+R\cdot i\omega\tau}}. \qquad (6.18)$$

Here, $R = K_H/K_L = (\omega_H/\omega_L)^2$; $\omega_L = \sqrt{K_L/m}$ is the circular frequency of the mass supported by both chambers; $\omega_H = \sqrt{K_H/m}$ is the circular frequency of the mass supported by the upper chamber only. If the stiffness of the diaphragm is neglected, $R = N + 1$, $N = V_2/V_1$. The values of $R$ and $\tau$ are the main design parameters of the two-chamber pneumatic isolator.

As seen from Eq. (6.18), for small viscous resistance ($\tau \to 0$) the transmissibility tends to that of an undamped oscillator with frequency $\omega_L$; for large viscous resistance ($\tau \to \infty$) the transmissibility tends to that of an undamped oscillator with frequency $\omega_H$. These two limiting transmissibility functions are shown in Fig. 6.22 by dashed lines. Figure 6.22 shows a series of transmissibility curves for isolators with the same value of $R$ and different values of $\tau$. These curves have one invariant point: the absolute value of transmissibility is the same for any value of $\tau$ at $\omega = \omega_{inv}$:

$$\omega_{inv} = \omega_L\sqrt{\frac{2R}{R+1}}, \quad |T(\omega_{inv})| = \frac{R+1}{R-1}.$$

The thin line in Fig. 6.22 shows the transmissibility curve with minimum resonance amplification possible for the given value of $R$.

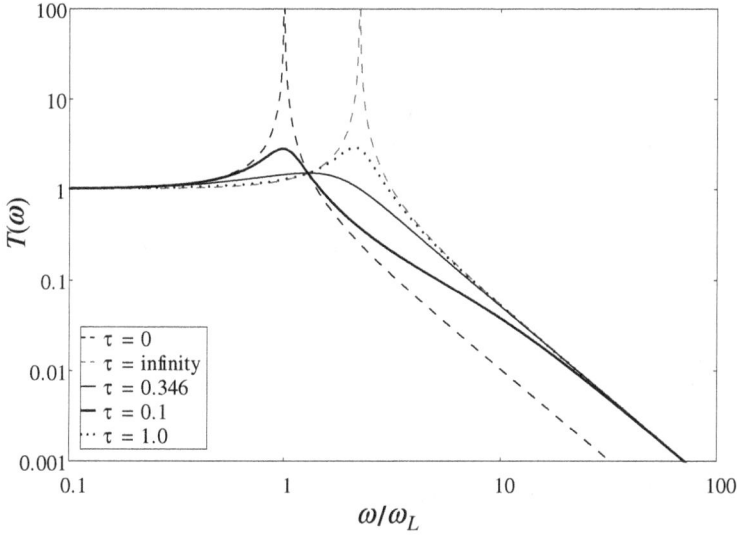

**Figure 6.22.** Transmissibility curves of a two-chamber pneumatic isolator for various levels of damping. $R = 5$. Time constant $\tau$ is normalized with respect to $(\omega_L)^{-1}$ [13].

This is achieved for a specific value of damping, $\tau = \tau_m$,

$$\tau_m = \frac{1}{\omega_L} \frac{\sqrt{R+1}}{R\sqrt{2}}. \tag{6.19}$$

Sometimes $\tau_m$ is considered an optimal value of damping. One could argue, however, that designs with lower damping have some advantages. Comparing the transmissibility shown by the thick solid line ($\tau = 0.1\omega_L^{-1}$) in Fig. 6.22 with the one shown by the thin line ($\tau = \tau_m = 0.346\omega_L^{-1}$), we note that the former one gives, in exchange for higher resonance amplification, higher performance in hard-to-achieve low-frequency isolation area just above $\omega_L$. The frequency band of vibration isolation expands to lower frequencies when $\tau$ decreases below $\tau_m$. On the other hand, it is clear from Fig. 6.22 that it makes no sense to increase $\tau$ beyond $\tau_m$ as shown by the dotted line since that would degrade the isolator performance at high frequencies while simultaneously increasing the resonance amplification. Therefore, $\tau_m$ should be considered the *maximum* recommended value of the normalized damping $\tau$.

The same isolator is normally expected to be used for supporting a range of loads. Due to the influence of atmospheric pressure and diaphragm stiffness, the value of $\omega_L$ is lower for higher loads and higher for lower loads. Therefore, according to Eq. (6.19), $\tau_m$ is higher for higher loads and lower for lower loads. The isolator should be designed so that the damping stays below the recommended maximum for the entire range of loads; therefore, it will appear less damped at high loads than at low loads. It should be noted that the reduction of $\omega_*$ (by increasing the volume) as a method of improving isolation is limited because it leads to reducing static and quasi-static stiffness.

As an example, Fig. 6.23 shows a family of transmissibility curves for one model of pneumatic vibration isolator under different loads.

**Exercise 6.1.** A pneumatic chamber has the form of a vertical cylinder 0.125 m in diameter. It is closed by a 5-kg piston. The distance from the bottom of a cylinder to the bottom surface of the piston is

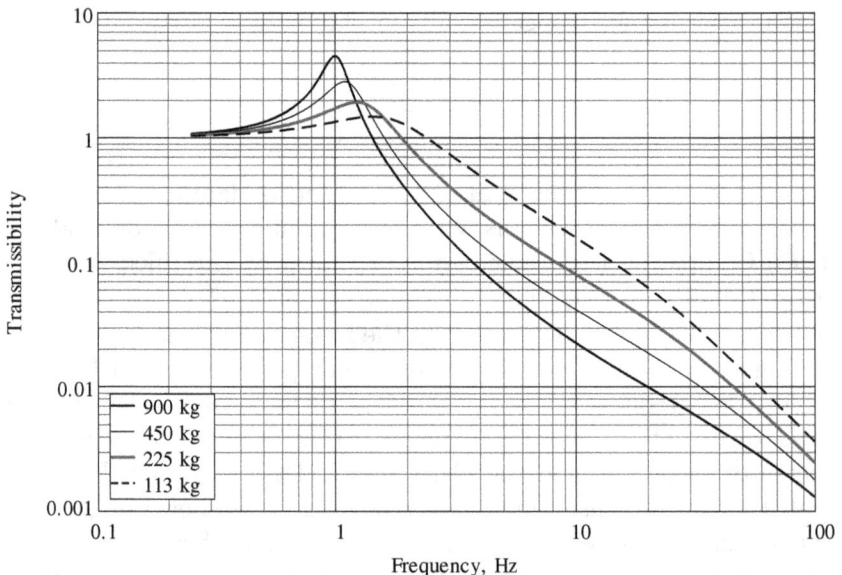

**Figure 6.23.** Transmissibilities of a pneumatic vibration isolator for several values of load per isolator.

0.25 m. What is the linear stiffness of the chamber? What is the natural frequency of small vertical oscillations of the piston supported by this air chamber?

**Answer:** According to Eqs. (6.7) and (6.8) with $p_{\text{atm}} = 101.3$ kPa,

$$K = 1.41 \frac{5 \cdot 9.8}{0.25} \left( 1 + \frac{101.3 \cdot 10^3 \cdot \pi \cdot 0.0625^2}{5 \cdot 9.8} \right) \approx 7288 \text{ N/m},$$

$$f_n = \frac{1}{2\pi} \sqrt{1.41 \frac{9.8}{0.25} \left( 1 + \frac{101.3 \cdot 10^3 \cdot \pi \cdot 0.0625^2}{5 \cdot 9.8} \right)} \approx 6.1 \text{ Hz}.$$

**Exercise 6.2.** A two-chamber pneumatic isolator has the following parameters: effective piston area 0.02 m², volume of upper chamber 0.002 m³, volume of surge chamber 0.008 m³. The isolator supports the load of 500 kg. The estimated stiffness of the diaphragm is 6 N/mm. What are the lowest (low frequency) and the highest (high frequency) values of its dynamic stiffness?

**Answer:**

$$K_L = \frac{\gamma p_0 A^2}{V_1 + V_2} + K_d$$

$$= \frac{1.41 \left( \frac{500 \cdot 9.8}{0.02} + 101.3 \cdot 10^3 \right) \cdot 0.02^2}{0.002 + 0.008} + 6 \cdot 10^3 \approx 25.5 \text{ N/mm},$$

$$K_H = \frac{\gamma p_0 A^2}{V_1} + K_d$$

$$= \frac{1.41 \left( \frac{500 \cdot 9.8}{0.02} + 101.3 \cdot 10^3 \right) \cdot 0.02^2}{0.002} + 6 \cdot 10^3 \approx 104 \text{ N/mm}.$$

Isolation in the horizontal direction is accomplished in state-of-the-art isolators by pendulum or post systems that are placed either inside the piston or outside the air chamber (see Fig. 6.24). The bottom part of the pendulum can be inserted in oil for damping. In some designs [16] an elastomeric element is placed in sequence with the air chamber. These arrangements provide linear response in the displacement range of interest. The resonance frequency of horizontal vibration is usually chosen approximately equal to that of vertical vibration unless application-specific requirements demand otherwise.

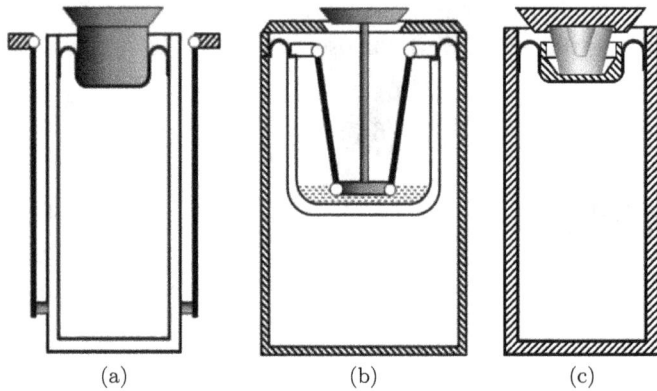

**Figure 6.24.** Pneumatic vibration isolators with horizontal motion provided by pendulums placed outside the pneumatic chamber (a) and inside the piston (b); elastomeric element placed in sequence with the pneumatic chamber (c).

## 6.5 Static and Dynamic Stability of Optical Table Systems

Vibration isolators are joined by the platform they support, so they do not work by themselves but form a mechanical system. It is especially true for pneumatic isolators with leveling valves since their air volumes can interact. This section, based on the paper [13], explains important static and dynamic properties of the systems of pneumatic isolators supporting a payload, with emphasis on static and dynamic stability.

### 6.5.1 *Load distribution in pneumatic vibration isolation systems*

When designing or specifying the isolation system, the first step is to make sure that the loads on individual isolators are within allowable limits. All isolators are rated to certain maximum loads. Some models also have minimum recommended loads. Most designs of isolators, pneumatic isolators in particular, are not intended to bear any negative (tensile) loads.

State-of-the-art pneumatic isolation systems are equipped with leveling valves that are used to set up the table initially in precise horizontal position and maintain this position against changes in the payload distribution or external perturbations. The system has always three leveling valves, notwithstanding the number of isolators. This means that in systems having more than three isolators some of the isolators must have their air volumes connected. When disturbed, they can freely trade air between themselves, so they do not offer any resistance to rotation about their common center. Standard ways of connecting the isolators in typical configurations are shown in Fig. 6.25. In a static position, the pressure is the same in each isolator of the group; therefore, each isolator bears the same load and has the same stiffness. A group of connected isolators is therefore statically equivalent to one isolator placed at their geometrical center with the stiffness equal to the sum of individual stiffnesses. Consequently, there are always three effective support points, and the system is statically determinate. In simple symmetric configurations, the load distribution between the isolators may be immediately evident. In the general case, a pneumatic system can support a complicated assembly of joined tables and payloads. The loads on each isolator should be estimated *a priori* to make sure that they are within the allowable limits.

To calculate the load on each isolator, first of all, three groups of connected isolators should be identified. Geometric centers of each group in the horizontal plane — denoted below as $(X_1, Y_1)$, $(X_2, Y_2)$, and $(X_3, Y_3)$ — are found by averaging $x$ and $y$ coordinates of

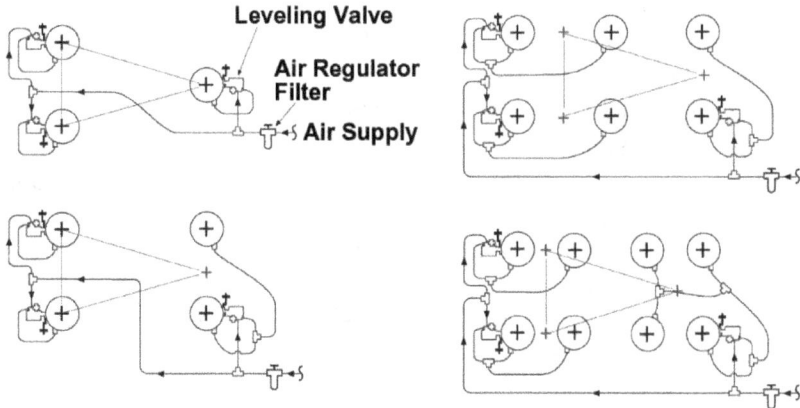

**Figure 6.25.** Schematics of pneumatic plumbing for systems of 3 isolators, 4 isolators, 6 isolators, and 8 isolators. Stability triangles are shown by dim lines. Adapted from [17]. Permission to use granted by Newport Corporation. All rights reserved.

individual isolators in the group. Then the position of the center of gravity, $(x_c, y_c)$, of the entire system, including tables and payloads, should be estimated. After that, the total loads on each group, $F_1$, $F_2$, and $F_3$, are determined by solving the equations of equilibrium,

$$F_1 + F_2 + F_3 = F, \quad F_1 x_1 + F_2 x_2 + F_3 x_3 = 0,$$
$$F_1 y_1 + F_2 y_2 + F_3 y_3 = 0,$$

where $F$ is the total weight of the system; $x_j$ and $y_j$, $j = 1, 2, 3$ — relative coordinates: $x_j = X_j - x_c$, $y_j = Y_j - y_c$. The solution can be written down explicitly:

$$F_1 = \frac{x_2 y_3 - x_3 y_2}{d} F, \quad F_2 = \frac{x_3 y_1 - x_1 y_3}{d} F, \quad F_3 = \frac{x_1 y_2 - x_2 y_1}{d} F,$$
$$d = (x_2 y_3 - x_3 y_2) + (x_3 y_1 - x_1 y_3) + (x_1 y_2 - x_2 y_1).$$

(6.20)

The load supported by each isolator is calculated by dividing the total load on the group by the number of isolators in the group, $n_j$, $j = 1, 2, 3 : F_j^0 = F_j / n_j$. The loads are positive if the CG falls within the triangle formed by the centers of the three groups of isolators with connected air volumes (see Fig. 6.25). Since the pneumatic isolators are not designed to withstand negative (tensile) loads, the CG should always stay within this "stability triangle" in the horizontal plane. In a well-arranged system, loads supported by isolators in each of the three groups are within the allowable range and are, preferably, close to each other.

The load distribution may be different in vibration-isolated systems based on springs or elastomers, as well as pneumatic isolators that are not self-leveling, so-called "fill-and-forget" isolators. Those systems can be statically indeterminate.

Another concern in setting up vibration isolators is the stability of the resulting system. Even if the load capacity of the isolators is sufficient, the equilibrium can be unstable if the center of gravity of the payload is too high above the support plane. The static stability analysis applies to vibration isolation systems of any kind. However, the main attention should be paid to pneumatic systems that are exceptionally soft. There are two types of potential instability in pneumatic vibration isolation systems. First, the system can become statically unstable because the center of gravity is too high. Second, the system can become dynamically unstable because the leveling valves have the gains set too high.

### 6.5.2 Static stability of vibration isolated systems with high centers of gravity

High-performance vibration isolation systems are characterized by very low stiffness of the isolators. State-of-the-art pneumatic vibration isolation systems can have natural frequencies of vertical motion as low as 1 Hz. That gives rise to potential static instability if payloads have high centers of gravity.

A general criterion for static stability states that a mechanical system in a position of equilibrium is stable if any small deviation from this position causes the potential energy to increase. Figure 6.26 illustrates this concept. The potential energy is a quadratic function (quadratic form) of small deviations of generalized coordinates from equilibrium values. The equilibrium is stable if this form is positive.

The following illustrative example helps understand static instability in a vibration-isolated system with high CG. Suppose that a rigid system is supported by springs symmetrically as shown in Fig. 6.27.

Assume, for sake of simplicity, that the motion is constrained to one vertical plane. Consider small rotation, $\varphi$, of this system around point $O$ which is the projection of the center of gravity (CG) on the isolators' plane. The horizontal motion can only add to the potential energy, as is the case, for example, of pendulum-type supports widely used with pneumatic isolators, so it can be set to zero for the static stability analysis. The change in the gravitational energy due to rotation $\varphi$ is

$$U_{\mathrm{g}} = -\frac{1}{2}mgH\varphi^2.$$

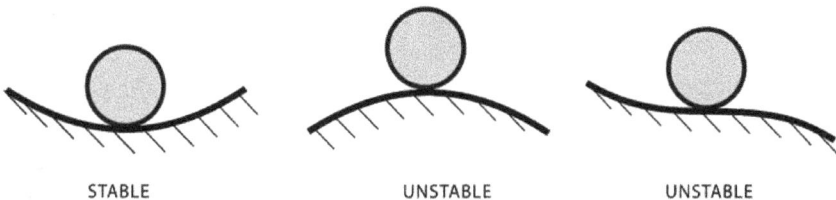

STABLE        UNSTABLE        UNSTABLE

**Figure 6.26.** The concept of static stability.

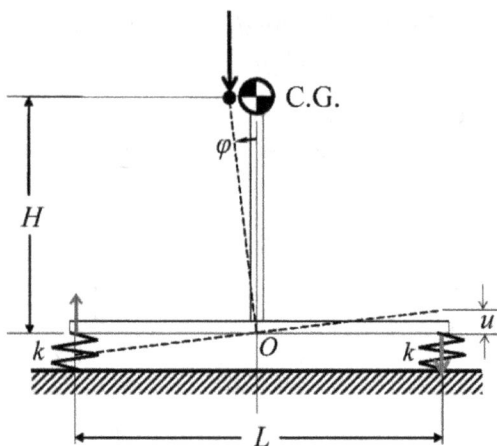

**Figure 6.27.**   A 2D model of a potentially statically unstable high-CG vibration isolated system.

The potential energy stored in the springs

$$U_e = 2 \cdot \frac{1}{2} k \left( \frac{L}{2} \varphi \right)^2.$$

The system is stable if the change in total potential energy, $U_g + U_e$, is positive for all non-zero values of $\varphi$, which leads to the following restriction on the height of the CG:

$$H < \frac{L^2}{2} \frac{k}{mg}. \tag{6.21}$$

For pneumatic isolators, the height $H$ is measured from the diaphragm level which defines the rotation centers of individual isolators.

In reality, possible tilt motions of the pneumatically isolated table are not restricted to one vertical plane, but the stability analysis in the general 3D case is performed by the same method. Consider a 3D rectangular coordinate system where the horizontal plane $(x, y)$ contains effective support points of isolators, and the vertical axis, $z$, passes through the CG. To analyze the static stability, calculate the changes in the potential energy due to small deviations from the equilibrium described by a small vertical motion, $w_0$, and a small rotation around an arbitrary horizontal axis by an angle $\varphi$ with components $\varphi_x$ and $\varphi_y$. As in the case of planar motion, only quadratic terms should be retained since the linear terms cancel out due to conditions of equilibrium. The change in the gravitational energy is

$$U_{\mathrm{g}} = -\frac{1}{2}mgH(\varphi_x^2 + \varphi_y^2), \tag{6.22}$$

where $m$ is the total mass of the system. The change in elastic energy of isolators is

$$U_{\mathrm{e}} = \frac{1}{2}\sum_{j=1}^{n} k_j w_j^2, \tag{6.23}$$

where $k_j$ are vertical stiffnesses and $w_j$ are vertical displacements of the isolators, $j = 1, \ldots, n$. For the rigid-body motion of the isolated platform, $w_j = w_0 + \varphi_y x_j - \varphi_x y_j$. Substituting these expressions in the formula for $U_{\mathrm{e}}$, and using the transformation

$$w_0 = w' - \frac{\sum k_j x_j}{\sum k_j}\varphi_y + \frac{\sum k_j y_j}{\sum k_j}\varphi_x,$$

the elastic energy can be written down in terms of rotation angles (assuming $w' = 0$) as

$$U_{\mathrm{e}} = \frac{1}{2}(I_x \varphi_x^2 - 2I_{xy}\varphi_x\varphi_y + I_y \varphi_y^2),$$

where

$$I_x = \sum k_j y_j^2 - \frac{\left(\sum k_j y_j\right)^2}{\sum k_j}, \quad I_y = \sum k_j x_j^2 - \frac{\left(\sum k_j x_j\right)^2}{\sum k_j},$$

$$I_{xy} = \sum k_j x_j y_j + \frac{\left(\sum k_j x_j\right)\left(\sum k_j y_j\right)}{\sum k_j}. \tag{6.24}$$

The system is statically stable if the quadratic form

$$U = \frac{1}{2}(I_x - mgH)\varphi_x^2 + \frac{1}{2}(I_y - mgH)\varphi_y^2 - I_{xy}\varphi_x\varphi_y$$

is positive, $U > 0$, for any $\varphi_x$ and $\varphi_y$. According to Sylvester's criterion [18], this form is positive if $I_x - mgH > 0$, $I_y - mgH > 0$, and $(I_x - mgH)(I_y - mgH) - I_{xy}^2 > 0$. This leads to the following general criterion of stability in terms of maximum allowed height of the center of gravity:

$$H < \frac{1}{mg}\left[\frac{I_x + I_y}{2} - \sqrt{\frac{(I_x - I_y)^2}{4} + I_{xy}^2}\right]. \qquad (6.25)$$

This relation gives the maximum allowable height of the CG as a function of its coordinates in the horizontal plane $(x_c, y_c)$.

In case of self-leveling pneumatic isolators, there are always three effective support points, $n = 3$, and stiffnesses, $k_j$, are total stiffnesses of groups of isolators with connected air volumes. The static stiffness of an individual isolator is, according to (6.7),

$$K = \gamma \frac{A^2}{V}(p_g + p_{atm}). \qquad (6.26)$$

The stiffness of the diaphragm is added to this expression when estimating the stiffness of the isolator under small displacements for calculating the vibration isolation. However, quasi-static disturbances may occur beyond the range of small displacements that usually spans just a few microns. In this case, the diaphragm "rolls" and does not provide substantial additional stiffness. For a safe estimate, the stiffness of the diaphragm should be neglected when calculating the static stability. As shown in the previous section, two-chamber pneumatic isolators behave as single-chamber with the total volume, $V_1 + V_2$, in the static and quasi-static regimes.

Since the gauge pressure, $p_g$, is defined by the load and the effective piston area, the total stiffness of each group of isolators according to Eq. (6.26) is given by

$$k_1 = \frac{F_1}{g}\omega_*^2 + k_a n_1, \quad k_2 = \frac{F_2}{g}\omega_*^2 + k_a n_2, \quad k_3 = \frac{F_3}{g}\omega_*^2 + k_a n_3, \quad (6.27)$$

where $\omega_* = \sqrt{\gamma g A/V}$ is the nominal circular natural frequency of the pneumatic chamber, $k_a = \gamma(A^2/V)p_{atm}$ is the stiffness defined by the atmospheric pressure, $F_k$ are the total loads on groups of isolators, $n_k$ is the number of isolators in the group. These values of stiffness should be used in expressions (6.24) to be substituted into the criterion (6.25).

Physically, there are two principal axes of rotation for any position of the CG in the horizontal plane: one with the maximum rotational stiffness, another with the minimum rotational stiffness. Stability requires that even for the minimum rotational stiffness the increase in elastic energy be greater than the decrease in gravitational energy for small rotations about the corresponding principal axis. The resulting stability domain for the CG is a dome-shaped area over the "stability triangle" in the horizontal plane that can be visually described as a combination of parabolic and pyramid shapes. Note that the height $H$ in the previous equations is measured from the diaphragm level of the isolator. It may be more convenient to use, instead, the height $H_b$ from the bottom surface of the isolated platform, $H_b = H - d_b$, where $d_b$ is the distance from the bottom surface to the diaphragm level. The examples of the stability domains for one model of a pneumatic isolator, calculated by Eq. (6.25), are shown in Figs. 6.28–6.31.

Simpler formulas that help understand the structure of these "stability surfaces" exist for the important case of the support system having a vertical plane of symmetry, with CG positioned in this plane

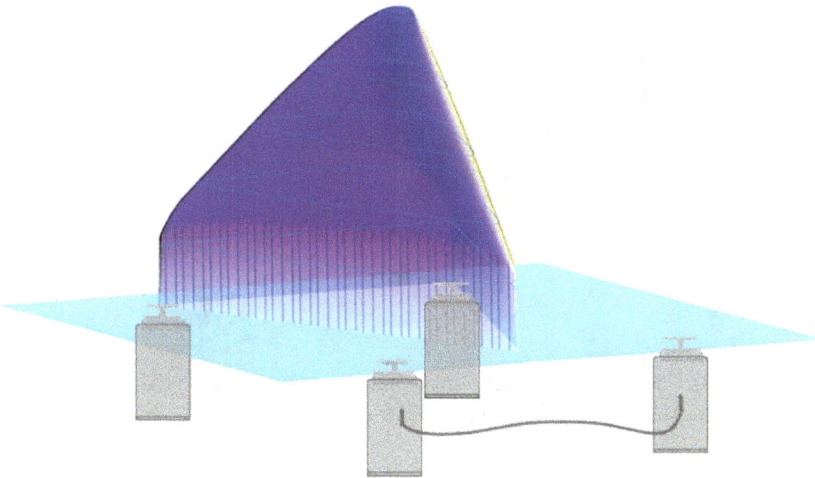

**Figure 6.28.** Stability domain for the CG of isolated payload. The light blue plane represents the bottom surface of the isolated platform (optical table). The isolators ($A = 0.017$ m$^2$, $V_1 = 0.42 \cdot 10^{-3}$ m$^3$, $V_2 = 0.91 \cdot 10^{-2}$ m$^3$, $d_b = 0.061$ m) form a rectangle 0.914 m $\times$ 1.321 m, two of them connected along the long side of the table. Isolators are drawn not to scale. Total load 1,820 kg. Maximum height $H_b = 0.657$ m over the bottom surface of the table. Restriction on maximum admissible load on each isolator is not taken into account.

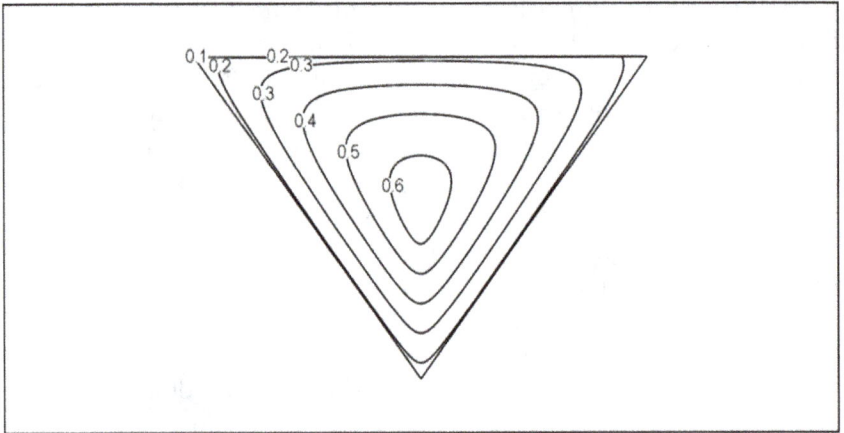

**Figure 6.29.**   Contour lines for Fig. 6.28. Height in m.

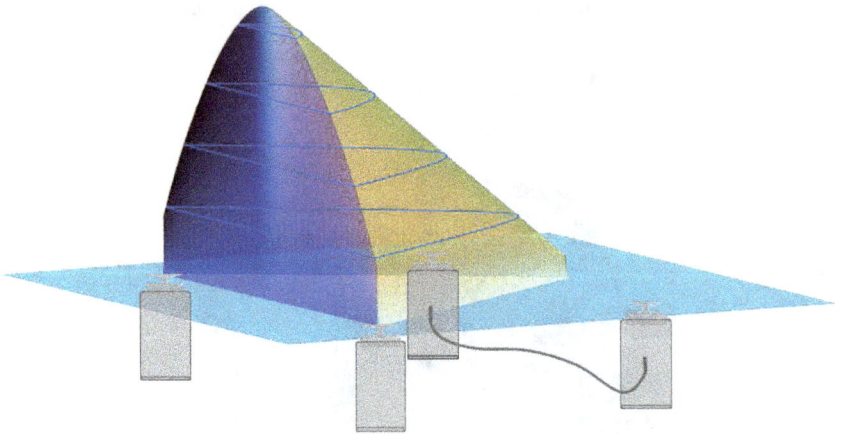

**Figure 6.30.**   Same as Fig. 6.28, but isolators connected along the short side of the platform. Maximum height $H_b = 0.556$ m.

of symmetry. This corresponds to the crests of the stability domains shown in Figs. 6.28–6.31. In this case one of the principal axes coincides with the axis of symmetry and the other is perpendicular to it. The maximum height of the CG over the axis of symmetry can be determined from the criterion (6.25) where $I_{xy} = 0$, so that the

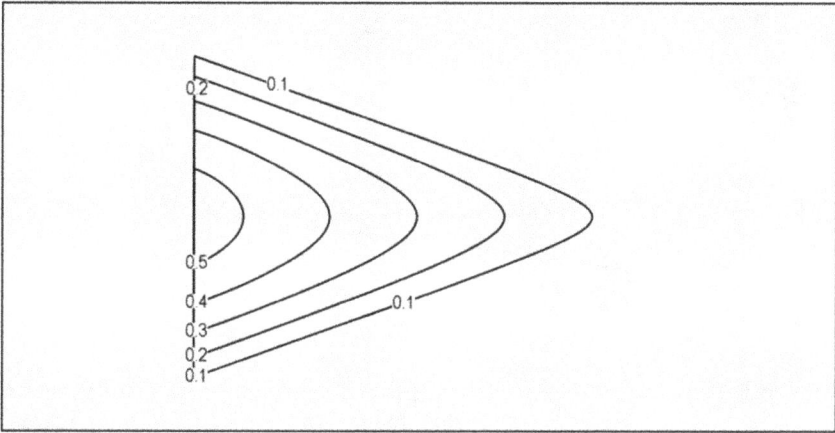

**Figure 6.31.** Contour lines for Fig. 6.30. Height in m.

equation becomes

$$H < \frac{1}{mg} \min(I_x, I_y),$$

or directly by calculating the balance of elastic and gravitational energies for small rotations around the axis of symmetry and the perpendicular axis. Either method leads to the following equations:

$$H(x) < \min \left\{ H_1(x), H_2(x) \right\}, \tag{6.28}$$

$$H_1(x) = \frac{\omega_*^2}{g} \frac{W^2}{4} \left( \frac{L-x}{L} + 2\frac{n_1}{n_t}r \right), \tag{6.29}$$

$$H_2(x) = \frac{\omega_*^2}{g} \left[ \frac{x(L-x)}{1+r} + \frac{r}{1+r} L \left( \frac{n_t - 2n_2}{n_t}x + \frac{n_2}{n_t}L + 2rL\frac{n_1 n_2}{n_t^2} \right) \right]. \tag{6.30}$$

Here $L$ is the height and $W$ is the base of the horizontal stability triangle, $n_t$ is the total number of isolators, $n_1 = n_3$ is the number of isolators connected together at each of two effective base support points, $n_2$ is the number of isolators connected together at the apex of this triangle. For example, Fig. 6.25 shows the following four cases: $n_t = 3$, $n_1 = n_2 = n_3 = 1$; $n_t = 4$, $n_1 = n_3 = 1$, $n_2 = 2$; $n_t = 6$, $n_1 = n_3 = n_2 = 2$; $n_t = 8$, $n_1 = n_3 = 2$, $n_2 = 4$. The value of $r$ describes the role of atmospheric pressure: $r = n_t \cdot A \cdot p_{\text{atm}}/F$ is the

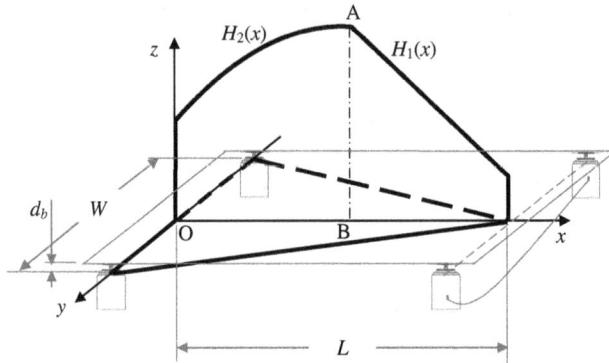

**Figure 6.32.** Stability domain for a symmetric configuration with CG placed in the plane of symmetry. Three-dimensional presentation of the stability domain for the same isolation system is shown in Fig. 6.28. $L = 0.914$ m, $W = 1.321$ m. OB $= 0.484$ m; AB $= 0.716$ m; $\max[H_2(x)] = 0.718$ m (from the diaphragm level). Isolators are drawn not to scale.

ratio of the atmospheric pressure to the average gauge pressure in the isolators' chambers. Figure 6.32 shows, as an example, the details of the crest of the stability domain of Figs. 6.28 and 6.29 as calculated from Eqs. (6.28)–(6.30). If the center of gravity is placed in the $(x, z)$ plane to the right of the line AB, the critical instability mode is the rotation around the $x$-axis, and the maximum height is defined by $H_1(x)$. If the center of gravity is in the $(x, z)$ plane to the left of AB, the critical instability mode is the rotation around the perpendicular axis, and the maximum height is defined by $H_2(x)$.

In reality, the maximum load on each isolator is limited. This restriction means that the CG of the entire system must stay close to the center of the table in the horizontal plane so that no individual isolator is overloaded. This reduces the allowable domain for the CG. Figures 6.33–6.36 show how stability domains of Figs. 6.28–6.31 are reduced by taking into account the maximum load of 8900 N per isolator.

In summary, the procedure for determining and adjusting the load distribution and static stability of a pneumatic isolation system is as follows. First, calculate the loads on individual isolators using Eqs. (6.20). Then calculate the stiffness of individual isolators and estimate the static stability using the criterion (6.25) or, in symmetric cases, (6.28)–(6.30). If the loads on the isolators are

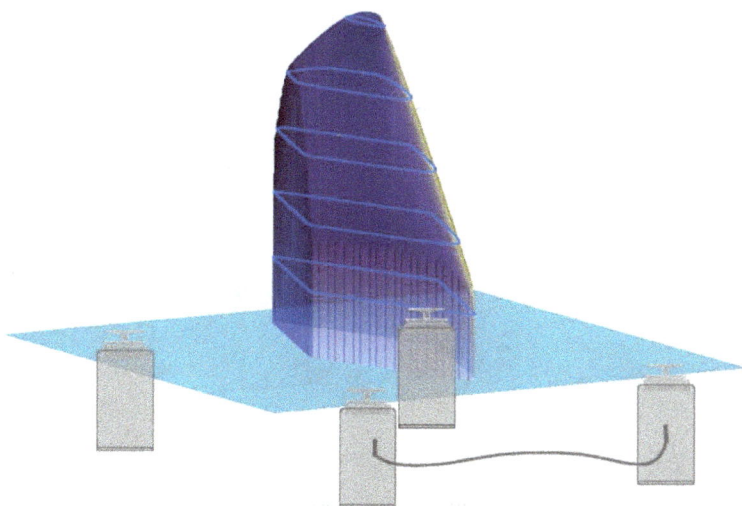

**Figure 6.33.** Same as Fig. 6.28, but the restriction of maximum admissible load on each isolator (8900 N) is taken into account.

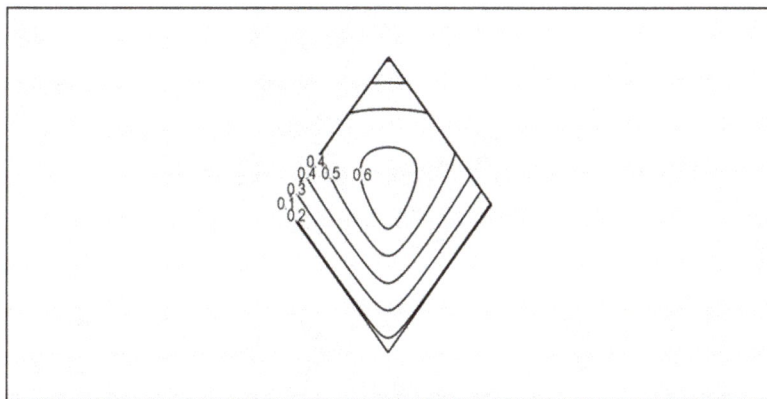

**Figure 6.34.** Contour lines for Fig. 6.33. Height in m.

outside the desired range, or the CG is too high outside the stability domain, the system should be modified by changing the air volume connection arrangement between isolators, changing the positions or number of the isolators. In some cases, recesses, outriggers, or cradle systems (Fig. 6.37) for isolators may be required for achieving static stability.

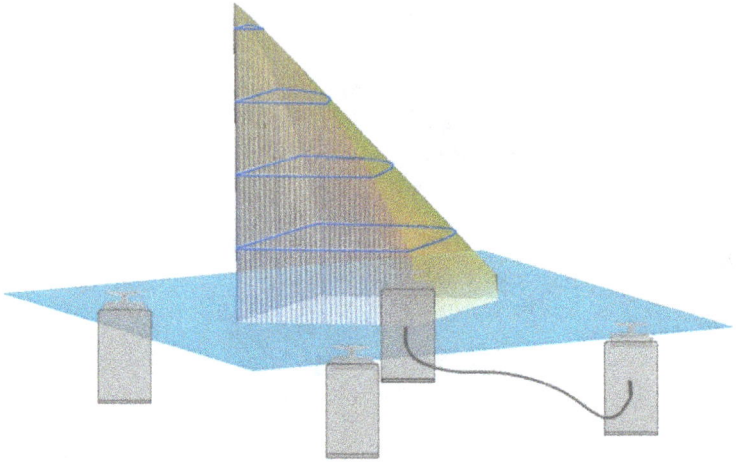

**Figure 6.35.** Same as Fig. 6.30, but the restriction on the maximum admissible load on each isolator (8900 N) is taken into account.

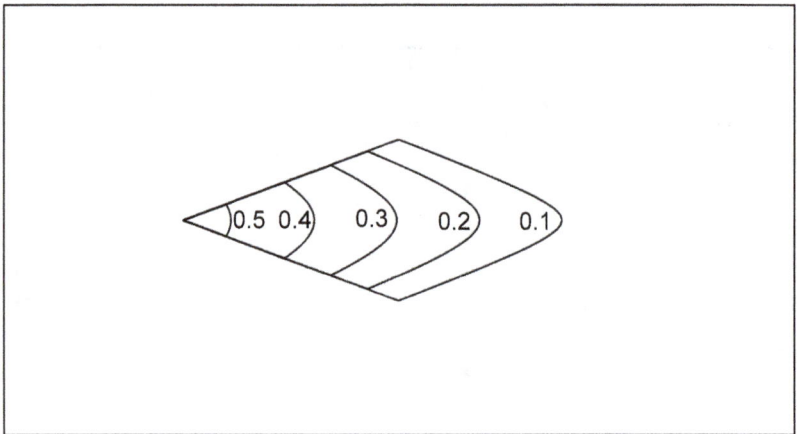

**Figure 6.36.** Contour lines for Fig. 6.35. Height in m.

**Exercise 6.3.** An isolated platform is supported by four pneumatic isolators that form a square with 1 m sides. Two of them, adjoining one side of the square, have their air volumes connected. The platform carries a load placed centrally so that the center of gravity of the platform and the payload is positioned exactly above the center of the square. The resonance frequency of vertical vibration is 1 Hz. If the payload is elevated so that the CG is 60 cm above the diaphragm level of the isolators, will the system be statically stable?

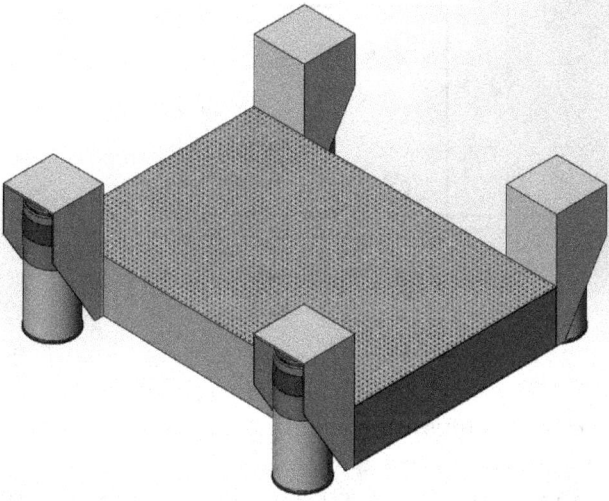

**Figure 6.37.** A cradle system improves static stability of the pneumatically isolated optical table.

**Solution.** This section gives three ways to solve for the static stability of the pneumatic isolation system. First, applicable universally to any configuration of isolators and payloads, is given by the criterion (6.25) with substituted expressions (6.24) and (6.27). Although it is not complicated (it is just an algebraic procedure that can be implemented in any mathematical software, even in an Excel worksheet), it is too cumbersome for this symmetric case; besides, it requires more detailed information about the volume of the chamber. The second way, described by Eqs. (6.28)–(6.30), applies to the systems where the support system has a plane of symmetry, and the CG is placed in this plane. Our case fits in this category. Assume first that the pressure in the chamber is much higher than the atmospheric pressure so that the parameter $r = n_t \cdot A \cdot p_{\text{atm}}/F$ in Eqs. (6.29) and (6.30) can be neglected. Then replace the nominal resonance frequency $\omega_*$ of the chamber with the observed natural frequency $\omega_0$. These two assumptions compensate for each other. So, Eqs. (6.28)–(6.30) simplify to

$$H(x) < \min\left\{H_1(x), H_2(x)\right\}, \quad H_1(x) = \frac{\omega_0^2}{g}\frac{W^2}{4}\left(\frac{L-x}{L}\right),$$

$$H_2(x) = \frac{\omega_0^2}{g}x(L-x). \tag{6.31}$$

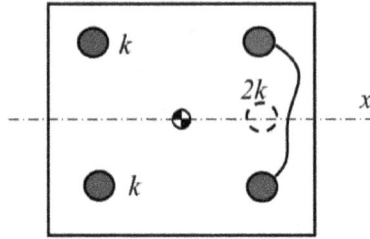

**Figure 6.38.** Illustration to Exercise 6.3.

Here $\omega_0 = 2\pi f_0$, $f_0 = 1$ Hz, $L = W = 1$ m, $x = L/2 = 0.5$ m, $g = 9.8$ m/s$^2$. Formulas (6.31) give $H_1 = 0.503$ m, $H_2 = 1.006$ m. So, the maximum admissible height is 50.3 cm, and the system with $H = 60$ cm is statically unstable.

Finally, the third and the simplest way is to use the solution for the 2D case (6.21). It is evident from symmetry that the critical rotation is around the $x$-axis (see Figs. 6.32 and 6.38). We should pay attention to the correct value of $k$. In our case, the mass $m$ is supported by four springs $k$. The load is the same on all 4 springs. Therefore, for the given $\omega_0$, $4k = m\omega_0^2$; $k = (1/4)m\omega_0^2$. Two pneumatic isolators are connected and therefore statically equivalent to one spring of double stiffness placed on the axis of rotation, therefore they do not contribute to the righting force. So, the correct solution is to use the formula (6.21) with $k = (1/4)m\omega_0^2$. The answer is $H < (L^2/8)(2\pi f_0)^2/g = 0.503$ m. Since the actual height is 60 cm, the system is statically unstable.

Note that in both solutions we neglected the contribution of the rolling diaphragm to the stiffness, so our estimate of the maximal stable value of $H$ is conservative.

### 6.5.3 *Gravitational terms and vibration transmissibility*

As the system approaches static instability, its lowest natural frequency tends to zero. This means that the performance analysis of a high-CG system must consider the gravitational components of the potential energy and, correspondingly, of the stiffness matrix, which is sometimes ignored in texts on vibration isolation. The performance of the isolation system can be characterized by its transmissibility.

Assuming the isolated platform is modeled as a rigid body, and the foundation supporting the isolators stays rigid in the frequency range under consideration, the transmissibility is a $6 \times 6$ matrix connecting the displacements of the origin and rotations of the platform, $(u, v, w, \varphi_x, \varphi_y, \varphi_z)^{\mathrm{T}}$, to the same values of the foundation. The system is characterized by a $6 \times 6$ inertia matrix $\mathbf{M}$ and a generalized frequency-dependent stiffness matrix $\mathbf{C}(\omega)$ that is defined by stiffness terms derived from the elastic energy (6.23) and the gravitational terms derived from the potential energy of gravity (6.22). Transmissibility between the generalized displacements of the foundation and the platform is expressed as the matrix (3.41) of Chapter 3.

The dependence of the transmissibility on the position of the CG is illustrated by the following example.

**Example 6.1.** As an example, consider again the system described in Figs. 6.28, 6.29, and 6.32, assuming that the CG is placed directly above the center of the platform. Vertical stiffnesses of the isolators are given by (6.26), the horizontal stiffnesses are approximated by $k_{\mathrm{hor}} = m\omega_{\mathrm{hor}}^2(1 + 0.3i\omega/\omega_{\mathrm{hor}})$, $\omega_{\mathrm{hor}} = 2\pi \cdot 1.5 \text{ s}^{-1}$. Assume that the mass is distributed symmetrically, so that $J_x = J_y = m\rho_c^2$, $J_{xy} = 0$. The stiffness of the diaphragm $K_d$ for this load was identified as $5.5 \text{ N/mm}$. Consider two different assumptions concerning the distribution of mass. In the first case, assume that the mass is concentrated around the center so that the radius of inertia is relatively small: $\rho_c = W/4$. In the second case, assume that the payload is spread around with a larger radius of inertia: $\rho_c = W/2$. Figure 6.39 plots the transmissibility in the horizontal $x$-direction from the floor to the center of gravity of the isolated platform. The graphs of Fig. 6.39 show, first, that the difference between vibration transmissibilities calculated with gravitational terms included and gravitational terms omitted can be very substantial. Second, high CG can, in fact, have a positive effect on vibration isolation — at the expense of lower stability margins.

### 6.5.4 *Self-leveling systems with mechanical feedback*

Pneumatic vibration isolators can have very low linear stiffness. So, even small changes of the load could lead to unacceptably large deviations in the position of the isolated platform. To avoid this, most

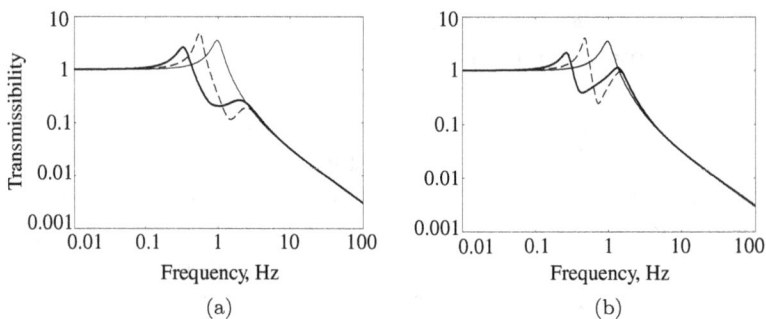

**Figure 6.39.** Thick line: transmissibility from the floor to the CG for symmetric payload on four isolators. Dashed line: the same calculated without gravitational terms. Thin line: nominal horizontal transmissibility of the isolators. (a) low moments of inertia, $\rho_c = W/4$. (b) high moments of inertia, $\rho_c = W/2$ [13].

of the state-of-the-art pneumatic vibration isolators are equipped with self-leveling valves ensuring a stationary position of the payload. These valves allow some air into the chamber if the arm of the valve shifts down and let some air out from the chamber if the arm shifts up.

Consider a single standalone two-chamber pneumatic isolator shown in Fig. 6.19(c) supporting a mass $m$ and equipped with a mechanical valve that allows a certain volume of air $q_v$ at a flow rate of $\dot{q}_v$ into the bottom chamber in response to the displacement of the payload:

$$\dot{q}_v = -Gu_1,$$

where $G$ is the gain of the valve. The mechanical model of the two-chamber isolator (6.12) must be augmented as follows:

$$P_1 = K_1(u_1 - u_2) + K_d u_1,$$
$$K_1(u_1 - u_2) - K_2(u_2 - u_v) = C_{12}\dot{u}_2, \qquad (6.32)$$
$$\dot{u}_v = -G_0 u_1,$$

where $u_v = q_v/A$, $G_0 = G/A$. One should have in mind that the re-leveling motion happens in the displacement range much wider than vibration amplitude typical for laboratory environments. The re-leveling valves usually have dead bands of tens to hundreds of $\mu$m. Therefore, for a safe analysis, we neglect the stiffness of the diaphragm, $K_d$, and the corresponding loss factor. Accordingly, the dynamic stiffness of the height-controlled isolator becomes

$$K_G(\omega) = \frac{K_L + K_H i\omega\tau + \frac{G_0}{i\omega}K_L}{1 + i\omega\tau}, \qquad (6.33)$$

where

$$K_L = K_1 K_2/(K_1 + K_2) = \gamma p_0 A^2/(V_1 + V_2), \quad K_H = K_1 = \gamma p_0 A^2/V_1.$$

If the isolator supports a payload modeled by the mass $m$, the dynamic process of re-leveling is described by the following equation in the frequency domain:

$$[-m\omega^2 + K_G(\omega)]u_1 = 0. \qquad (6.34)$$

To analyze the stability, one must research the position of the solutions of the characteristic equation $-m\omega^2 + K_G(\omega) = 0$ in terms of the Laplace variable $s = i\omega$:

$$ms^2(1 + \tau \cdot s) + K_L + K_H \cdot \tau \cdot s + \frac{G_0 K_L}{s} = 0. \qquad (6.35)$$

The system is asymptotically stable if all the roots reside in the left half-plane of the complex $s$ plane. This leads, via the Routh–Hurwitz criterion, to the following condition of stability [13]:

$$G_0 < \tau(\omega_H^2 - \omega_L^2), \qquad (6.36)$$

where $\omega_L^2 = K_L/m$, $\omega_H^2 = K_H/m$. The criterion can be put in a non-dimensional form in several ways:

$$\frac{G_0}{\omega_L} < \tau\omega_L N; \quad \frac{G_0}{\omega_L} < \frac{C_{12}}{m\omega_L}\left(\frac{N}{N+1}\right)^2; \quad \frac{G_0}{\omega_L} < 2\zeta\left(\frac{N}{N+1}\right)^2. \qquad (6.37)$$

Here $N = R - 1$; since the stiffness of the diaphragm is neglected, $N = K_1/K_2 = V_2/V_1$; the damping ratio, $\zeta$, is defined by analogy with a simple damped oscillator: $C_{12} = 2m\omega_L\zeta$.

Stability analysis of a multi-support system of pneumatic isolators presents a more complicated task. Stability depends on the height of the center of gravity and the moments of inertia of the supported load. Vertical, horizontal, and rotational degrees of freedom must be considered. The condition of stability is derived from dynamic equations of the payload (represented as a rigid body together with the table) supported by the isolators, with feedback control included. Without going into the full mathematical derivation [13], consider the results summarized in Fig. 6.40. This graph relates, again, to the system described in Figs. 6.28, 6.29, and 6.35, assuming that the CG is placed directly above the center of the platform. Figure 6.40 plots the maximum stable gain as function of the CG height for the two cases: low moments of inertia of the payload (including the platform) $\rho_c = W/4$ and high moments of inertia $\rho_c = W/2$. The circle and square symbols denote results obtained by a simplified method, applying Eq. (6.37) for the single-degree-of-freedom case with the lowest natural frequency of the undamped uncontrolled system in place of $\omega_L$. The system is unstable for all values of gain, including zero, if the height of the CG is beyond that determined by static stability. Therefore, a statically unstable system cannot be stabilized by operating the leveling valves. For low-profile payloads with small

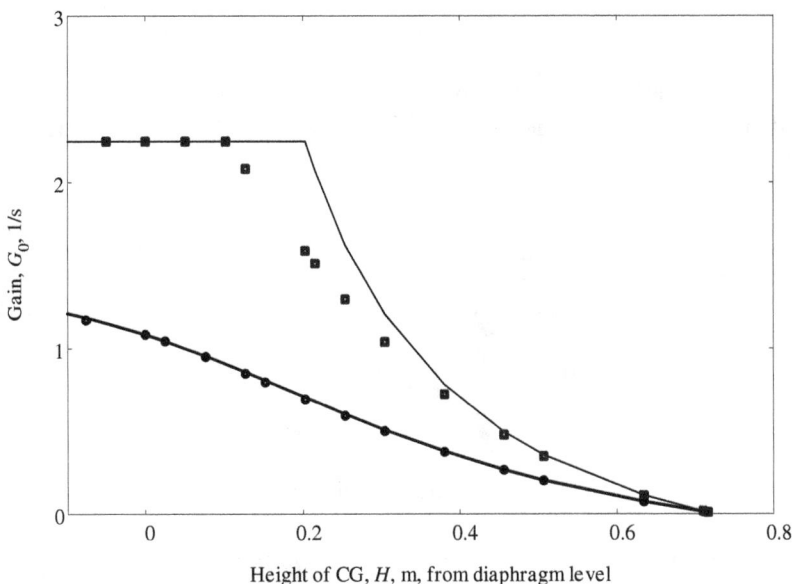

Height of CG, $H$, m, from diaphragm level

**Figure 6.40.** Stability limits for a symmetric payload on four isolators with the CG placed directly above the center [13]. Total load 1,820 kg. The maximum height is 0.718 m from the diaphragm level as in the static case. Thick line — high moments of inertia; thin line — low moments of inertia. The maximum gain equals 2.242 s$^{-1}$ in accordance with Eqs. (6.36)–(6.37). Symbols show results of a simplified analysis.

radii of inertia, the maximum gain is the same as that determined from Eq. (6.37). This is the case when the lowest natural frequency corresponds to the vertical mode of vibration. In other cases, rotational (rocking) modes have lower natural frequencies, and the maximum gain is lower. In most cases, the simplified method yields the results close to that of the full analysis. The deviations are due to horizontal damping that is not taken into account by the simplified procedure.

Stability limits such as those shown in Fig. 6.40 can be used directly for design of precision height control systems with proportional valves. If the valves are not proportional, as it often happens in practice, these data give a qualitative understanding of the relationship between the system design parameters, such as payload geometry and placement of isolators, and the stable area of valve operation. Practically, if the leveling valves are used, it is recommended to reduce the maximum allowable CG height obtained from static analysis by 15–20% for heavy payloads concentrated near the middle of the table, 20–30% if the payloads are spread over the surface of the platform.

In summary, there are three distinct operation ranges of the pneumatic isolators in terms of displacement amplitudes. The displacement range of precision vibration isolation in typical laboratory settings covers only a few micrometers. The diaphragm is not rolling, and its full stiffness should be included in the stiffness of the isolator when estimating the vibration transmissibility. The leveling valves are closed within their dead bands. In the displacement range from a few micrometers up to the dead band of leveling valves (usually tens or hundreds of micrometers) the diaphragm is rolling, which provides diminishing resistance to the motion of the piston. This displacement range is critical for the static stability analysis; the stiffness of the diaphragm should be neglected for a safe design. Finally, the range beyond the dead band of the re-leveling valves is important for the dynamic stability analysis of the feedback system.

# References

1. E.E. Ungar and C.W. Dietrich, High-frequency vibration isolation, *Journal of Sound and Vibration*, vol. 4, no. 2, pp. 224–24, 1966.
2. E.I. Rivin, *Passive Vibration Isolation*, ASME, New York, 2003, 432 p.

3. Ya.G. Panovko and I.I. Gubanova, *Stability and Oscillation of Elastic Systems: Modern Concepts, Paradoxes and Errors.* NASA, Washington, DC, 1973, pp. 197–201.

4. I.A. Karnovsky and E. Lebed, *Theory of Vibration Protection,* Springer, Switzerland, 2016, pp. 26–27.

5. Smalley Steel Ring Company, *Engineering and Parts Catalog,* 2015.

6. https://www.newport.com/f/vip320-vibe-mechanical-vibration-isolator-platforms. Accessed on March 1, 2021.

7. P. Alabuzhev, A. Gritchin, L. Kim, G. Migirenko, V. Chon, and P. Stepanov, *Vibration Protecting and Measuring Systems with Quasi-Zero Stiffness,* CRC Press, New York, 1989, 100 p.

8. D.L. Platus, Negative-stiffness-mechanism vibration isolation systems. *SPIE Proceedings Volume 3786, Optomechanical Engineering and Vibration Control,* 1999, pp. 98–105.

9. J.D. Ferry, *Viscoelastic Properties of Polymers,* 3rd edn., Wiley, New York, 1980.

10. C.M. Harris and C.E. Crede (eds.), *Shock and Vibration Handbook,* vol. 2, 1st edn., McGraw Hill, New York, 1961, p. 35–19.

11. Passive vibration isolator with profiled supports. US Patent 6394407, 2002.

12. *Outgassing Data for Selecting Spacecraft Materials,* NASA Ref. Pub. 1124, rev. 4, 1997; see also https://outgassing.nasa.gov/.

13. V.M. Ryaboy, Static and dynamic stability of pneumatic vibration isolators and systems of isolators, *Journal of Sound and Vibration,* vol. 333, pp. 31–51, 2014.

14. M. Heertjes and N. van de Wouw, Nonlinear dynamics and control of a pneumatic vibration isolator, *Transactions of the ASME, Journal of Vibration and Acoustics,* vol. 128, no. 4, pp. 439–448, 2006.

15. C. Erin, B. Wilson, and J. Zapfe, An improved model of a pneumatic vibration isolator: theory and experiment, *Journal of Sound and Vibration,* vol. 218, no. 1, pp. 81–101, 1998.

16. Pneumatic vibration isolator utilizing an elastomeric element for isolation and attenuation of horizontal vibration. US Patent 6619611, 2003.

17. *Newport Resource,* Catalog, Newport Corporation, 2004 and later editions.

18. L. Meirovich, *Computational Methods in Structural Dynamics,* Sijthhoff & Noordhoff, The Netherlands, 1980, p. 95.

Chapter 7

# Vibration Damping

Vibration damping is one of the three existing methods of vibration control, the other two being reduction of vibration in its source and vibration isolation (treated in Chapter 6). On the other hand, vibration damping, or dissipation of mechanical energy, is arguably the most important phenomenon one needs to understand in order to successfully apply vibration control. This chapter talks about damping in general terms and specifically in application to vibration-isolated platforms such as optical tables.

## 7.1  Methods of Damping and Types of Vibration Dampers

The models and examples considered in the previous chapters (see specifically Section 3.2 of Chapter 3) show that damping manifests itself in two ways: first, it reduces the resonance amplification; second, it causes attenuation of transient vibrations.

There are two distinct types of vibration damping usually encountered in vibration damping devices. They are sometimes called viscous damping and dry, or structural, damping. In the first case the dissipation of mechanical energy is achieved by a viscous fluid, usually metered in narrow passages or confined spaces. A dashpot is an example of a viscous damping device; a pendulum immersed in oil used for horizontal isolation in a pneumatic vibration isolator (see Fig. 6.24(b) of Chapter 6) is another example. In the second case, the damping is achieved by energy dissipation in a solid. Elastomeric materials exhibiting viscoelastic behavior are most often used for this purpose.

The dissipated mechanical energy in an elementary volume of material during a cycle of vibration, $E$, is proportional to the loss factor times the potential energy stored in this elementary volume, which is, in turn, the shear modulus times shear deformation squared:

$$E \propto \eta \frac{1}{2} G \gamma^2. \tag{7.1}$$

Here $G$ is the shear modulus, $\gamma$ is the shear deformation, and $\eta$ is the loss factor. The volume deformation is usually not accompanied by substantial dissipation of mechanical energy.

There are several known methods of applying damping treatments [1]. The first type of treatment is the so-called free-layer damping treatment. It is normally applied to thin-walled structures (Fig. 7.1). The layer of highly damped material is simply attached to the surface of the main structure. It is often very easy to implement. However, it is not very effective. As noticed above, the elastic energy is dissipated when the material is dynamically deformed in shear. In the free layer, the deformation is mostly extension-compression with very little shear, so the dissipation is relatively low. This shortcoming is corrected in a more sophisticated type of treatment, called constrained-layer damping treatment (Fig. 7.2).

Here another layer, much stiffer than the damping layer — typically comparable in stiffness to the main structure — is attached over the damping layer. This material restricts the extension of the top surface of the damping layer, thereby causing high shear deformation

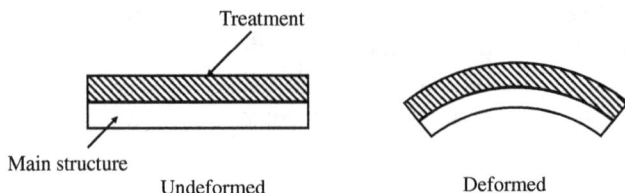

**Figure 7.1.**   Free-layer damping treatment.

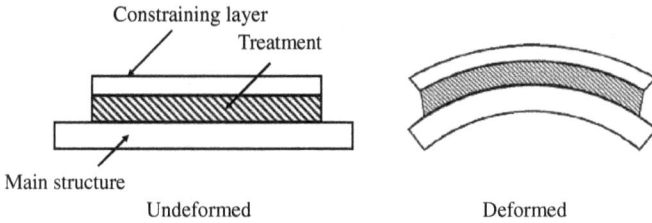

**Figure 7.2.** Constrained-layer damping treatment.

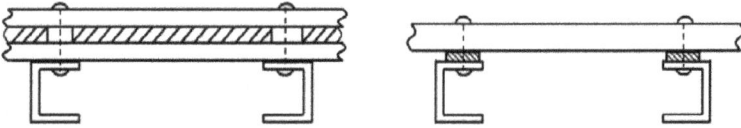

**Figure 7.3.** Integral damping treatments. Dashed elements represent visco-elastic materials.

that produces, according to Eq. (7.1), significant loss of energy. Constrained layer damping found wide applications in vibration control; detailed guidance and examples of use can be found in [1–3].

Viscoelastic materials used for damping treatments include elastomers such as polyurethanes, natural and synthetic rubbers, and silicones in form of cured polymers, pressure-sensitive adhesives, damping tapes, and baked-on or spray-on mastics.

Integral damping treatments (Fig. 7.3) are treatments that are added during the manufacturing or assembly process so that they become part of the structure itself. They include integrated laminated sheets of damping material, or damping material cast in the forms of shims or washers, or otherwise introduced into interfaces between parts of the structure.

Damping links are damping devices or inserts joining parts of the structure that experience large relative motion as shown, e.g., for a frame structure in Fig. 7.4. This technique is widely used in seismic construction. We'll consider an example of a damping link applied to an optical breadboard later in this chapter.

Damping devices relying on dry friction, e.g., those based on wire mesh, or visco-plastic substances such as magneto-rheological fluids, do not find wide application in precision vibration control of optomechanical systems because of their non-linear properties including a

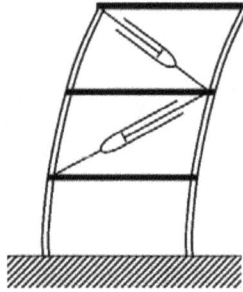

**Figure 7.4.** Damping links in a frame structure.

dead band: there is a range of small forces that do not cause any relative motion across the device, therefore no dissipation of mechanical energy occurs. This is usually not acceptable in precision vibration control.

## 7.2 Principle of Action and Optimum Parameters of Tuned Mass Dampers

Tuned mass dampers, or dynamic vibration absorbers, are considered the most effective means of damping [4]. Their principle of action is based on the phenomenon of antiresonance. To reduce the resonance vibration of a given mechanical system (the *primary* system), a *secondary* system is attached to it with its partial resonance frequency (*tuning frequency*) close to the resonance frequency of the primary system. Although the principle of action is simple, the tuning frequency and the level of damping in the secondary system should be chosen carefully to reach the optimal damping performance.

There is extensive literature on dynamic vibration absorbers. The fundamental monograph by Korenev and Reznikov [5] summarizes the development up to the publication date. With growing number of applications, renewed interest in optimization of dynamic absorbers under various criteria and assumptions was addressed in the works of Asami, Nishihara, and Baz [6, 7] as well as Zilletti, Elliott, and Rustighi [8]; design procedures for dynamic absorbers applied to multi-modal objects were proposed by Krenk and Høgsberg [9, 10].

Basic theory of the dynamic absorber can be derived using a two-degree-of-freedom model shown in Fig. 7.5. Here the oscillator

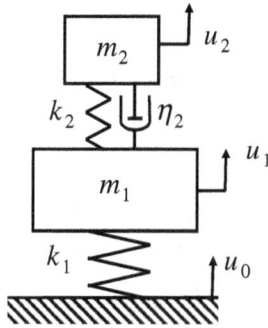

**Figure 7.5.** Two-DOF model of the system with dynamic vibration absorber.

modeled by the mass $m_1$ and stiffness $k_1$, represents the primary low-damped system whose resonance vibration needs to be damped, and the secondary system modeled by the mass $m_2$, stiffness $k_2$, and loss factor $\eta_2$, represents the dynamic vibration absorber.

The goal is to reduce the resonance vibration of the primary system. To achieve this, the partial resonance frequency of the vibration absorber is made close to the natural frequency of the primary system. We know from Section 3.2 of Chapter 3 that this secondary system will exhibit antiresonance, that is, its dynamic stiffness will be high at this frequency, which will likely reduce the resonance amplitude of $m_1$; we also know that the secondary mass, $m_2$, will exhibit large dynamic displacement relative to $m_1$, which should lead to dissipation of energy. To choose specific parameters that would make it work, consider the vibration transmissibility from the foundation to $m_1$ as a criterion. This function is derived from the two-degree-of-freedom model as explained in Section 3.3 of Chapter 3. Using the matrix notation, the equation of motion is given by Eq. (3.28) of Chapter 3, where

$$\mathbf{M} = \begin{pmatrix} m_1 & 0 \\ 0 & m_2 \end{pmatrix}, \quad \mathbf{C} = \begin{pmatrix} k_1 + k_2 & -k_2 \\ -k_2 & k_2 \end{pmatrix}, \tag{7.2}$$

and the transmissibility can be calculated directly using formulas (3.29) or (3.30) of Chapter 3, where $k_2(1 + i\eta_2)$ must be substituted instead of $k_2$ to account for the (frequency-independent) damping. Direct calculation gives the transmissibility as a complex function of frequency:

$$T(\omega) = \frac{k_1[k_2(1+i\eta_2) - m_2\omega^2]}{[k_1 + k_2(1+i\eta_2) - m_1\omega^2][k_2(1+i\eta_2) - m_2\omega^2] - [k_2(1+i\eta_2)]^2}.$$
$$(7.3)$$

Here the damping of the primary system is considered negligible, which is true in most cases requiring dynamic absorbers for damping. Damping of the primary system can be included by considering complex stiffness in place of $k_1$.

The appropriate model of damping in the dynamic absorber depends on the material used for damping. A highly damped viscoelastic elastomer is often chosen so that its loss factor is close to maximum in the vicinity of the target frequency; in this case, the frequency dependence of the loss factor is relatively flat in the frequency range of interest, and the model with frequency-independent loss factor serves well for approximate analysis. If the viscous fluid, such as oil, is employed for damping the elastic element of the secondary mass, then the viscous damping model is more suitable. In this case, the complex stiffness including the damping proportional to velocity should be used in place of $k_2$ in (7.2), and the transmissibility takes the form

$$T(\omega) =$$
$$\frac{k_1[k_2 + 2i\zeta_2\sqrt{m_2 k_2}\,\omega - m_2\omega^2]}{[k_1 + k_2 + 2i\zeta_2\sqrt{m_2 k_2}\,\omega - m_1\omega^2][k_2 + 2i\zeta_2\sqrt{m_2 k_2}\,\omega - m_2\omega^2] - [k_2 + 2i\zeta_2\sqrt{m_2 k_2}\,\omega]^2},$$
$$(7.4)$$

where $\zeta_2$ is the damping ratio of the secondary system.

In other cases, general models of complex frequency-dependent stiffness (see Section 3.2 of Chapter 3) can be used for describing the elastic element of the isolator.

**Exercise 7.1.** Derive the modulus of the transmissibility of the system with dynamic absorber modeled by a frequency-independent loss factor (7.3) in the real form.

**Answer:**

$$|T(\omega)| = \frac{k_1\sqrt{(k_2 - m_2\omega^2) + k_2^2\eta_2^2}}{\sqrt{[(k_1 + k_2 - m_1\omega^2)(k_2 - m_2\omega^2) - k_2^2]^2 + k_2^2\eta_2^2(k_1 - (m_1 + m_2)\omega^2)^2}}.$$

This function is plotted in Fig. 7.6 for different values of the loss factor.

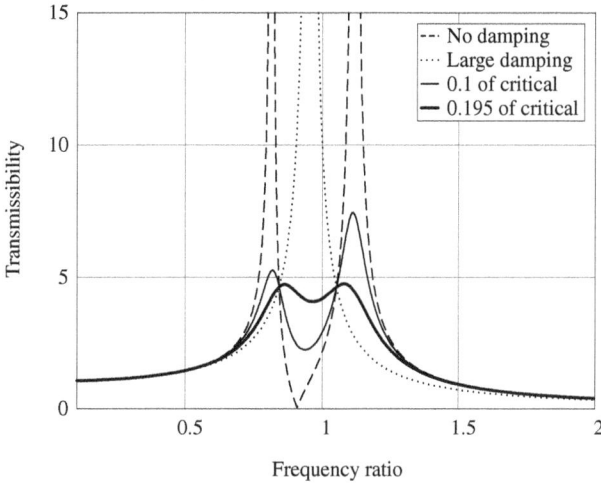

**Figure 7.6.** Vibration transmissibility of the oscillator with dynamic vibration absorber (Fig. 7.5) assuming frequency-independent damping. Mass ratio $\mu = 0.1$. The curves correspond to optimal frequency tuning by formula (7.5) and different values of the loss factor. The thick solid line shows the optimal transmissibility according to Eq. (7.6).

**Exercise 7.2.** Derive the modulus of the transmissibility of the system with a viscously damped dynamic absorber (7.4) in real form.

**Answer:**

$$|T(\omega)| = \frac{k_1 \sqrt{(k_2 - m_2\omega^2) + 4m_2 k_2 \zeta_2^2 \omega^2}}{\sqrt{\left[(k_1 + k_2 - m_1\omega^2)(k_2 - m_2\omega^2) - k_2^2\right]^2 + 4m_2 k_2 \zeta_2^2 \omega^2 (k_1 - (m_1 + m_2)\omega^2)^2}}.$$

This function is plotted in Fig. 7.7 for different values of damping ratio.

Optimization of the dynamic absorber seeks the optimal relationship between the natural frequency of the main system $\omega_0 = \sqrt{k_1/m_1}$ on one hand, and the parameters of the absorber: undamped natural frequency $\omega_a = \sqrt{k_2/m_2}$ and the loss factor $\eta_2$ on the other hand. The mass ratio

$$\mu = m_2/m_1,$$

is the main parameter of the system. Various criteria of optimization can be considered [3–6]; usually, minimization of the maximum value of transmissibility is a good criterion. Note that the same functions (7.3) and (7.4) give the ratios of dynamic compliances at the point

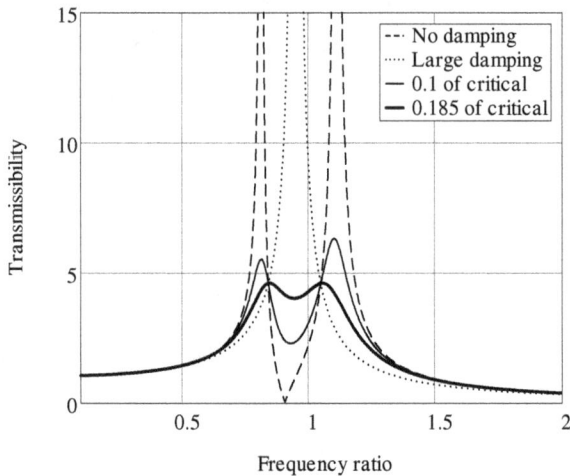

**Figure 7.7.** Vibration transmissibility of the oscillator with dynamic vibration absorber (Fig. 7.5) assuming viscous damping. Mass ratio $\mu = 0.1$. The curves correspond to optimal frequency tuning by formula (7.5) and different values of the loss factor. The thick solid line shows the optimal transmissibility according to Eq. (7.7).

$m_1$ to the static compliance $(1/k_1)$, so minimization of the transmissibility is at the same time minimization of the dynamic compliance.

For both viscous and hysteretic models of damping, the tuning frequency ratio

$$\frac{\omega_a}{\omega_0} = \frac{1}{1+\mu}, \tag{7.5}$$

is recommended as optimal. The optimal levels of damping are estimated by

$$(\eta_2)_{\text{opt}} = \sqrt{\frac{(3+2\mu)\mu}{2+\mu}}, \tag{7.6}$$

in case of frequency-independent damping, and

$$(\zeta_2)_{\text{opt}} = \sqrt{\frac{3\mu}{8(1+\mu)}}, \tag{7.7}$$

in case of viscous damping. The optimization of the damping in the dynamic absorber is illustrated in Figs. 7.6 and 7.7. As shown in

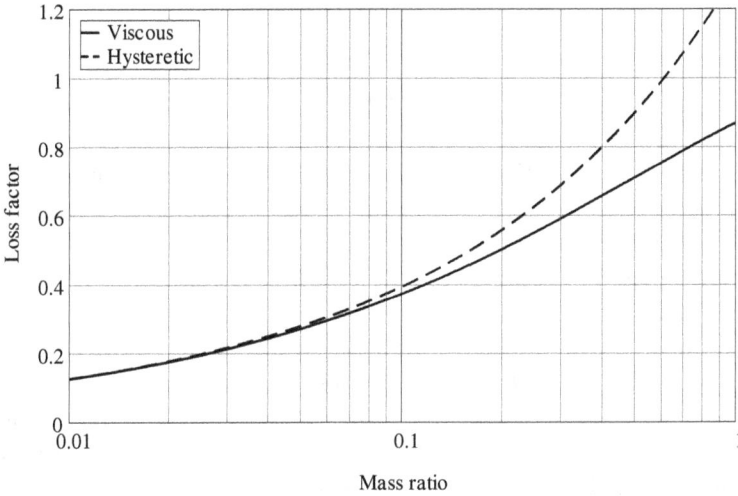

**Figure 7.8.** Optimal loss factor in the elastic element of the dynamic absorber as function of the mass ratio, $\mu$. Solid line — viscous damping, dashed line — hysteretic damping.

Fig. 7.8, the difference in optimal levels of damping (assuming $\eta_2 = 2\zeta_2$) is small for moderate values of the mass ratio $\mu$.

In the best case, the maximum transmissibility can be reduced, for moderate values of $\mu$, approximately to

$$T_{\text{opt}} = \sqrt{1 + \frac{2}{\mu}}. \tag{7.8}$$

This means that dynamic absorbers with larger mass can provide better performance.

Based on these considerations, the following general approach to the design of a dynamic absorber can be proposed. First, estimate the available mass budget ($\mu$). Then, use the graphs of Fig. 7.8 for guidance in choosing the material providing the optimum amount of damping. Finally, create elastic elements producing the required resonance frequency of the absorber. Use Eq. (7.5) if the primary system can be approximated by a simple oscillator; use *ad hoc* optimization for more complicated models of the primary system.

**Exercise 7.3.** A rigid elastically supported bench weighing 100 kg exhibits excessive resonance vibration at 40 Hz. The goal is to make the resonance amplification of floor vibration not higher than 12 dB.

Define the main parameters of vibration absorber to achieve this goal.

**Solution.** According to Table 2.2 of Chapter 2, 12 dB correspond to $T = 4$. Substituting this into Eq. (7.8) leads to $\mu = 2/(T^2 - 1) = 0.133$, which means that the moving mass of the absorber should be at least $m_2 = \mu m_1 = 100$ kg·0.133 $= 13.3$ kg. According to (7.5), $f_a/f_0 = \omega_a/\omega_0 = 1/(1 + \mu)$; $f_0 = 40$ Hz, therefore, the tuning frequency $f_a = 40$ Hz/1.133 $\approx 35$ Hz. If the materials with largely frequency-independent loss factors are used, the resonance frequency of the absorber tested independently is the same (see Exercise 3.4 of Chapter 3). Total dynamic stiffness of the resilient elements of the absorber is $k_2 = (2\pi f_a)^2 m_2 = 6.6 \cdot 10^5$ N. The optimal loss factor, according to Eq. (7.6) and Fig. 7.8, is $\eta = 0.45$. This level of damping is feasible for polyurethanes and silicone rubbers. A spring–elastomer combination is possible as described in the next section.

Tuning dampers for continuous structures can be based on the modal representation of their local dynamic compliance at the attachment point, see Eq. (3.37) of Chapter 3:

$$C(x, \omega) = \frac{u(x)}{F(x)} = \sum_{k=1}^{\infty} \frac{1}{m_k(x)} \frac{1}{\omega_k^2(1 + i\eta_k) - \omega^2}.$$

In this case, $m_k$ can be considered as an effective mass of the main structure, and the mass ratio, defined by

$$\mu_k = (m_a)_k/m_k, \tag{7.9}$$

is used for tuning the damper. A more precise methodology, taking into account the influence of other modes, is described in [9, 10].

**Exercise 7.4.** The effective mass of the first flexural mode of a structure, measured at a certain location, is 100 kg. The resonance frequency is 150 Hz. Somebody suggested installing at this location a tuned mass damper with a moving mass of 10 kg. What should be the tuning frequency, $f_a$, of this damper?

**Solution:** According to Eq. (7.9), $\mu = 10$ kg/100 kg $= 0.1$. According to (7.5), $f_a/f_0 = \omega_a/\omega_0 = 1/(1 + \mu)$; $f_0 = 150$ Hz, therefore, $f_a = 150$ Hz/1.1 $\approx 136$ Hz.

Practical design case studies of the dynamic absorbers can be found in [2]. One design of vibration absorber suitable for vibration-isolated platforms such as optical tables is described in the next section.

## 7.3 Dynamic Properties and Damping of Vibration-Isolated Structures: Application to Optical Tables

As explained in Chapter 6, a vibration-isolated structure includes two main components: soft isolators and a stiff isolated platform. The platform, designed to carry vibration-sensitive optomechanical applications, should be as stiff as possible to minimize misalignment of optical paths. It may take a form of a honeycomb optical table, a granite slab, a metal frame, a stiff casing, or another structure perceived as sturdy and stable. An optical table is the most common implementation of an isolated platform. Ideally, the platform should behave as an absolutely rigid body; unfortunately, absolutely rigid bodies do not exist: any structure deviates from the rigid behavior at its own resonance frequencies. Resonance vibrations of the platform carrying optomechanical equipment can be excited by acoustical inputs, residual vibrations coming from the floor through isolators, and atmospheric turbulence. These vibrations must be mitigated by dissipation of mechanical energy, or damping.

Vibration-isolated structures are characterized by multi-degree-of-freedom or continuous models described in Sections 3.4 and 3.5 of Chapter 3. When talking about the dynamics of optical tables, or any isolated platforms, we need to distinguish between two different phenomena: rigid-body vibration and flexural vibration. The rigid-body oscillation was described in Section 3.4 and specifically in Exercise 3.8 of Chapter 3. In a well-designed system the frequencies of these oscillations, often described as "bouncing" on the isolators, should be much lower than the frequencies related to the elastic, or flexural, deformation of the tabletop. Flexural deformations are defined by the natural modes of the table. Significant natural frequencies for typical laboratory sizes of the tables are clustered between 100 and 500 Hz. So, rigid-body vibration and flexural vibration are separated in the frequency domain and there is practically no inter-action between the two. The first is controlled by stiffness and internal damping of isolators; the second is controlled by structural and damping properties of the tabletop itself. In the intermediate frequency range, the entire system behaves largely as a freely floating rigid body. This was illustrated by the beam model in Example 3.5 of Chapter 3.

Dynamic properties of an isolated platform are traditionally described by its dynamic compliance, that is, the ratio of dynamic displacement to dynamic force as function of frequency (see Section 3.2 of Chapter 3). Figure 7.9 shows an example of dynamic compliance measured near a corner of a 305 mm thick optical table. Low-frequency range covering the rigid-body (bouncing on isolators) modes is usually not shown on these diagrams. It is related to technical difficulties of covering the whole frequency range in one test.

The dashed line in Fig. 7.9 represents the reaction that would be recorded if the platform were absolutely rigid and free. It is a straight line in the log-log scale with the slope of 40 dB (100 times)

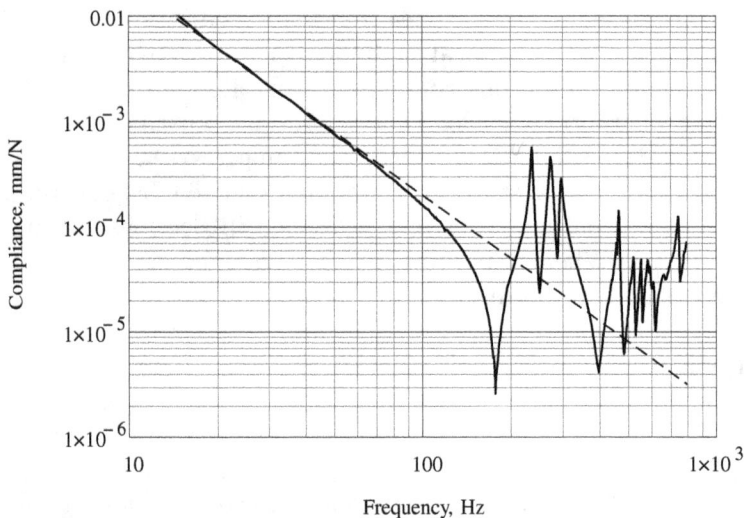

**Figure 7.9.** Example of measured dynamic compliance curve of an optical table. The dashed line describes the ideal rigid-body response. Rigid-body and resonance frequency ranges are clearly seen.

per decade, meaning the ratio of displacement to force inversely proportional to frequency squared in accordance with Newton's second law. The diagram shows large deviations from rigid-body lines and high vibration amplitudes at the resonance frequencies. This is due to relatively low internal damping in the table structure.

Figure 7.10 shows a typical design of an optical table. Stiff lightweight honeycomb core supports steel facesheets, which ensures high static and dynamic stiffness and high stiffness-to-mass ratio. Sealed hole tiles or other means of sealing the array of threaded holes prevent corrosion due to possible spills and facilitate the necessary cleaning. All elements are held together with a rigid epoxy or welding. The downside of the high stiffness of the optical benches is the lack of high integral damping. The loss factor of steel is below 0.001; epoxy and tile material can have a loss factor up to 0.02–0.05. The side panels in some designs are made of composite materials with a loss factor of about 0.1; other designs use metal side panels with lower damping. Together these properties lead to the modal loss factor of the whole structure not higher than 0.01–0.02. Therefore, additional means of damping are introduced in high-end optical tables. The tuned dynamic absorber described in the previous section is the most effective of these means.

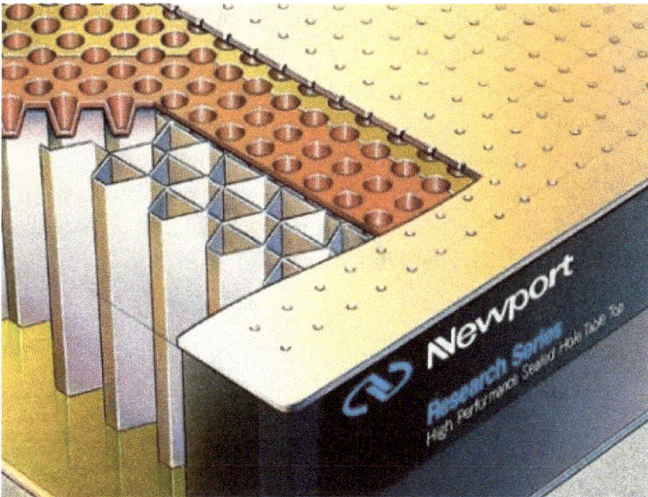

**Figure 7.10.** State-of-the-art design of a honeycomb optical table. Courtesy of Newport Corporation. All rights reserved.

Estimating the dynamic performance of an isolated platform, we note that two factors are important for minimizing the dynamic misalignment. First, the "rigid-body" frequency range of dynamic response should be made as wide as possible, meaning that the resonance frequencies should be as high as possible. Second, the resonance amplification should be made as low as possible, meaning that the associated loss factors should be made as high as possible.

**Example 7.1. Model of a pneumatically isolated platform.** To gain more insight into the dynamics of vibration-isolated platforms, let us return to the example considered in Chapter 3 (Example 3.5), but this time assume, instead of constant-rate springs, a more realistic model of vibration isolator, namely the model of a two-chamber pneumatic isolator discussed in Chapter 6 (Fig. 6.19(c)), characterized by the complex frequency-dependent stiffness (6.15). This example was originally developed as part of Newport Corporation's technical note.

The analytical solutions will allow for illustration of principal aspects of vibration isolation design, such as the role of structural and non-structural mass and bending stiffness of the platform. We'll consider two cases of excitation: linear motion of the foundation with the displacement amplitude $u_0$, Fig. 7.11(a), modeling the vibration transmissibility from the floor, and force excitation with amplitude $F$, Fig. 7.11(b), modeling effects of acoustical, atmospheric or on-board sources, as well as the standard dynamic compliance (hammer) test. Applying forces and elastic supports at the end of the beam allows for simpler analytical solutions obtained just by employing proper boundary conditions in each case.

Assume, for sake of simplicity, that the pneumatic isolators provide the constant natural frequency exactly (not approximately as described in Section 6.4 of Chapter 6) so that the complex stiffness

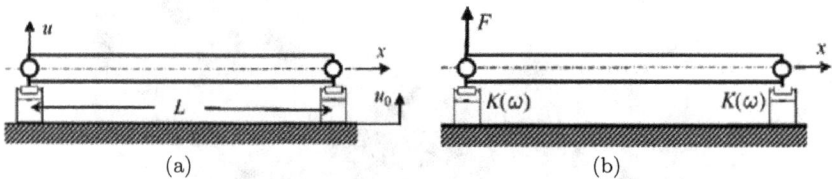

**Figure 7.11.** Model of a pneumatically isolated platform (Example 7.1). (a) Kinematic excitation; (b) force excitation.

(6.15) takes the form:

$$K(\omega) = K_L \frac{1 + R \cdot i\omega\tau}{1 + i\omega\tau}, \tag{7.10}$$

where $K_L = m \cdot (2\pi f_0)^2/2$, $f_0 = 1$ Hz as in Example 3.5; $R = K_H/K_L$. We assume in this example $R = 16$, $\tau = 0.02$ s, which is close to parameters of common pneumatic vibration isolators.

The frequency response of the beam can be calculated the same way as in Example 3.5 of Chapter 3, as a solution to the differential equation for the complex amplitude (3.46),

$$\frac{d^4 u(x)}{dx^4} - \kappa^4 u(x) = 0, \tag{7.11}$$

where

$$\kappa^4 = \frac{m}{L} \frac{\omega^2}{EI}.$$

As in Example 3.5 of Chapter 3, we adjust the bending stiffness of the beam, $EI$, so that the first natural frequency of bending vibration is 200 Hz, typical for laboratory-size optical tables.

Frequency-dependent stiffness of the elastic supports (7.10) dictates frequency-dependent boundary conditions for Eq. (7.11). They can be written down by substituting (7.10) in place of stiffness, $k$, into the boundary conditions of Example 3.5 of Chapter 3:

$$u''(0) = 0; \quad u''(L) = 0, \tag{7.12}$$

and

$$EIu'''(0) = -K(\omega)(u(0) - u_0); EIu'''(L) = K(\omega)(u(L) - u_0), \tag{7.13}$$

in case of kinematic excitation of the foundation with the amplitude $u_0$, Fig. 7.11(a), or

$$EIu'''(0) = -K(\omega)u(0) + F; \quad EIu'''(L) = K(\omega)u(L), \tag{7.14}$$

in case of force excitation $F$, Fig. 7.11(b). Further calculations of the complex amplitude $u(x)$ are completely analogous to Example 3.5

of Chapter 3 (see Eqs. (3.79) and (3.80)) using frequency-dependent stiffness ratio

$$k_r(\omega) = \frac{K(\omega)L^3}{EI}.$$

To avoid infinite amplitudes at flexural resonances, damping with a loss factor $\eta = 0.01$ was introduced into the structure by substituting complex value $EI(1+i\eta)$ in place of $EI$.

Vibration transmissibilities for the kinematic excitation from the foundation are calculated using the boundary conditions (7.13). The results are plotted in Fig. 7.12 along with the corresponding rigid-body solution shown by the dashed line. The graphs show that under the kinematic excitation the transmissibility $u/u_0$ rolls off the same way as for a rigid body (single degree of freedom) at frequencies up to the vicinity of the first resonance frequency of flexural vibrations. At that frequency, a large deviation from the rigid-body behavior occurs.

The analytical model allows for a more detailed analysis, calculating the misalignment of portions of the beam and relative displacement of points of the beam (Fig. 7.13).

The results of this analysis, shown in Fig. 7.14, reveal that even in the rigid-body zone small quasi-static deformations occur that do not roll-off with frequency. These deformations are caused by inertia forces generated by slow oscillations. The misalignment increases sharply at the flexural resonance: relative displacement reaches almost twice the absolute resonance response. Due to the

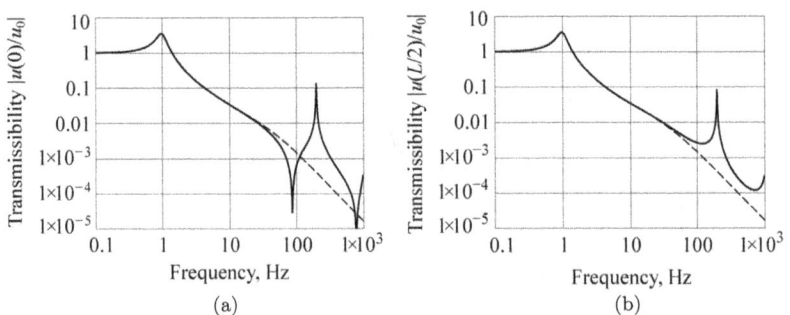

**Figure 7.12.** Example 7.1: Transmissibilities of the pneumatically isolated beam to the edge of the beam (a) and the center of the beam (b). Thick line — full model, dashed line — rigid beam. Only symmetric modes are excited by symmetric kinematic excitation, Fig. 7.11(a).

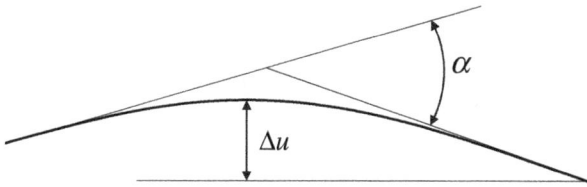

**Figure 7.13.** Relative displacement, $\Delta u$, and misalignment, $\alpha$.

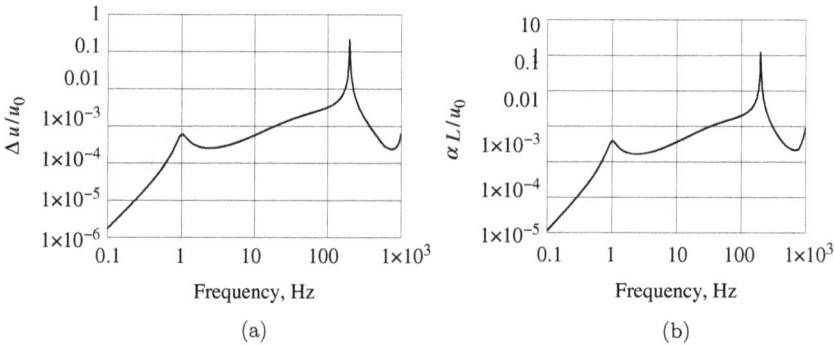

(a)                                  (b)

**Figure 7.14.** Example 7.1: (a) Maximum relative displacement between points of the pneumatically isolated beam, $\Delta u$, and (b) maximum misalignment between areas of the beam, $\alpha$, caused by the kinematic excitation $u_0$ from the foundation, Fig. 7.11(a). All functions are shown in non-dimensional form.

symmetric nature of the excitation, only symmetric vibration modes (see Fig. 3.31 of Chapter 3) are excited. Misalignment and relative displacement behave very similarly with respect to frequency.

In case of force excitation, Fig. 7.11(b), the boundary conditions (7.14) lead to the dynamic compliance at the excitation point and transfer functions to the middle of the beam plotted in Fig. 7.15. The "rigid body line" in that case is a straight line with a slope $1/f^2$ (40 dB per decade) corresponding to Newton's second law (acceleration proportional to force). The graph shows large deviations from the rigid-body response at the first and second bending modes of the beam (see Fig. 3.31 of Chapter 3).

Calculating the relative displacement of points of the beam due to flexural vibrations requires subtraction of the rigid-body motion; misalignment can be obtained by direct differentiation of the deflection. The results, plotted in Fig. 7.16, show, again, that, even in the rigid-body zone, small quasi-static deformations occur that do

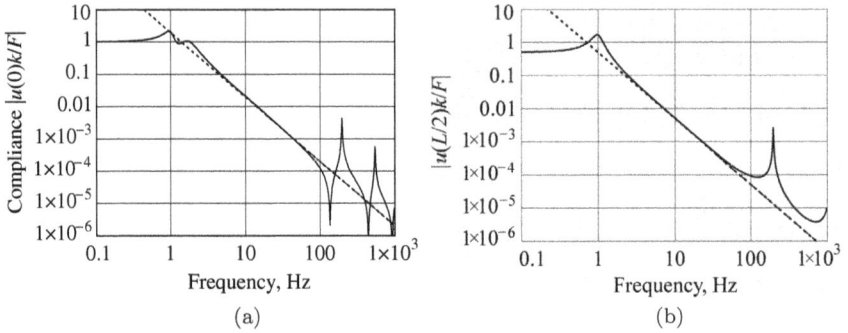

**Figure 7.15.** Example 7.1: Dynamic compliance at the end of the beam (a), and transfer functions from the force at the end of the beam to the displacement in the middle of the beam (b). Solid line — beam supported by pneumatic isolators, dashed line — rigid beam of equivalent mass, dotted line — ideal free rigid body. All functions are referred to the static compliance, $1/k$.

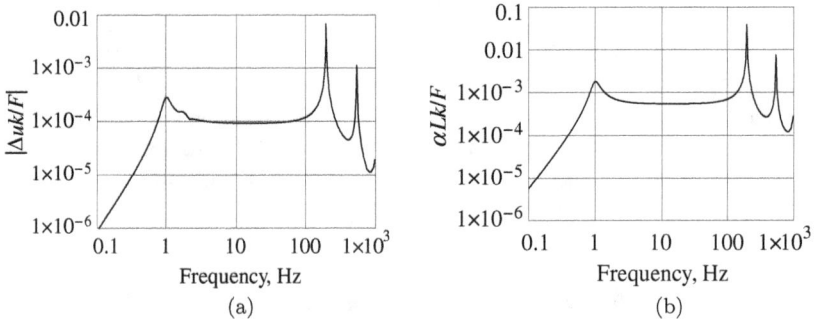

**Figure 7.16.** Example 7.1: (a) Maximum relative flexural displacement between points of the pneumatically isolated beam, $\Delta u$, (b) maximum misalignment between two areas of the beam, $\alpha$, caused by force excitation, Fig. 7.11(b), per unit force. All functions are reduced to non-dimensional form by referring to the static compliance, $1/k$.

not roll-off with frequency. These deformations are caused by inertia forces generated by slow oscillations. The misalignment increases at the flexural resonances: relative displacement reaches about twice the absolute resonance response at the first bending mode (200 Hz) and the second bending mode (551 Hz).

It is common practice to plot the transmissibility in a low-frequency range encompassing isolation resonances and part of the rigid-body frequency range, and plot the dynamic compliance at higher frequencies, encompassing the rigid-body range and resonance frequencies. This is related to the limitations of generally adopted experimental techniques. The analytical model considered in this example is free of these limitations and allows us to review simultaneously all aspects of the dynamic behavior of the isolated platform in a wide frequency range.

To analyze the influence of non-structural mass, suppose we double the mass of the beam without increasing the bending stiffness. The stiffness of the pneumatic isolators adjusts to keep the main "bouncing" frequency the same at 1 Hz. Figure 7.17 shows the resulting vibration transmissibility compared to the base case. The frequency range of the rigid-body behavior is reduced; the flexural resonance frequency becomes lower; the peak resonance response is increased; more resonance frequencies enter the depicted frequency range.

The effect of increasing the non-structural mass on relative motion and misalignment is illustrated in Fig. 7.18. Both relative motion and misalignment increase with increased mass.

Figure 7.19 shows the effect of doubling the mass on the dynamic compliance under the force excitation. For correct comparison, both functions are reduced to non-dimensional form by referring to the static compliance in the base case, $1/k$. The rigid-body line shifts down; on the other hand, the frequency range of rigid-body behavior contracts, and more resonances enter the depicted frequency range.

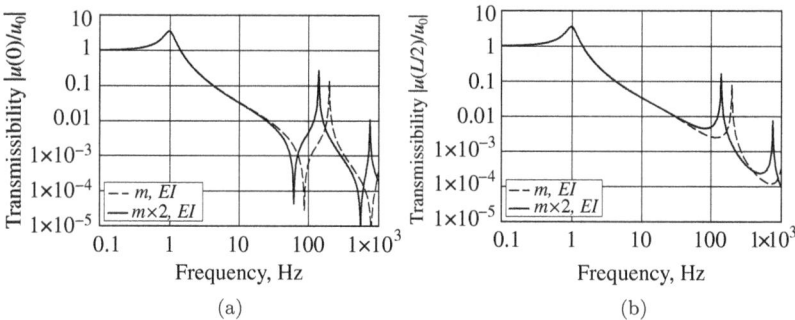

(a)                                         (b)

**Figure 7.17.** Example 7.1: Effect of increased non-structural mass on the transmissibility of the pneumatically isolated beam under the kinematic excitation $u_0$ from the floor. Resonance amplitudes increase; flexural resonance frequencies decrease; "rigid body" frequency range contracts; more resonances enter the given frequency span.

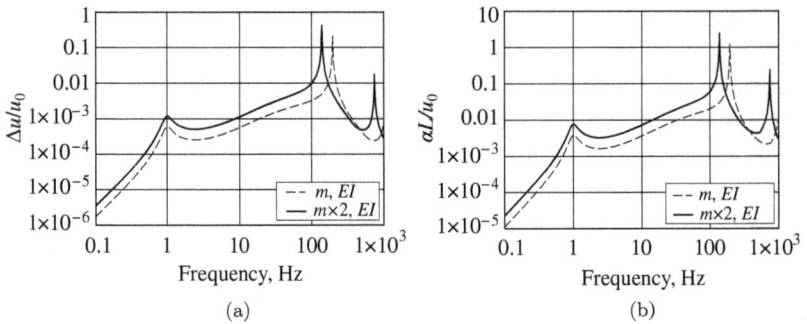

**Figure 7.18.** Example 7.1: Adding non-structural mass leads to increased relative displacement, $\Delta u$, and misalignment, $\alpha$, of the pneumatically isolated beam under kinematic excitation $u_0$ from the floor; the resonance peaks increase.

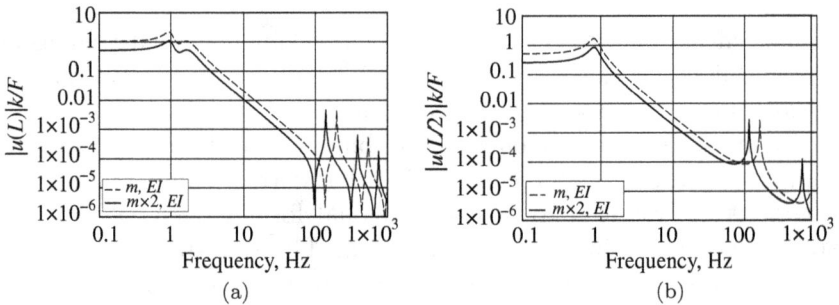

**Figure 7.19.** Example 7.1: Effect of increased non-structural mass on the dynamic compliance at the edge of the beam and vibrations in the middle of the beam under force excitation $F$. Flexural resonance frequencies decrease; "rigid body" range contracts and more resonances appear while the rigid-body line moves down; resonance amplitudes stay the same. All functions are reduced to non-dimensional form by referring to $1/k$.

Under the force excitation, the relative motion and the misalignment in the rigid-body zone stay the same with added non-structural mass, but the rigid-body zone itself contracts, and more resonances may appear in the given frequency range. It is illustrated in Fig. 7.20.

The behavior of the misalignment and relative motion in response to changing the non-structural mass has a simple physical explanation. Maximum relative motion of points in a structure is proportional to the motion relative to the center of mass (CG). To describe the motion of the structure in the coordinate system tied to the CG

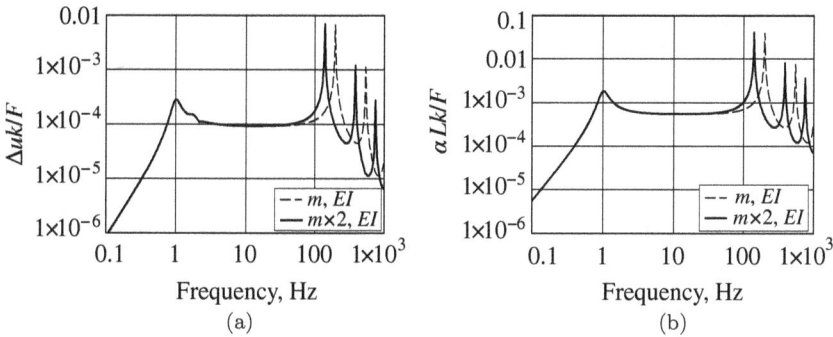

**Figure 7.20.** Example 7.1: Effect of increased non-structural mass on the maximum relative flexural displacement between points of the pneumatically isolated beam, $\Delta u$, and maximum misalignment between two areas of the beam, $\alpha$, caused by force excitation $F$, Fig. 7.11(b), per unit force. Both functions are reduced to non-dimensional form by referring to the static compliance in the base case, $1/k$.

we must apply the inertia forces to the structure. Inertia force acting on any material particle equals its mass times acceleration of the coordinate system. So, the inertia forces, and therefore elastic displacements, are proportional to the mass times acceleration of the CG. Acceleration equals displacement times $(2\pi f)^2$ in the frequency domain. The motion of the CG is the same as the rigid-body motion. Under the force excitation (Figs. 7.19 and 7.20), rigid-body displacement rolls-off like $1/f^2$, that is why the graphs of relative flexural motion and misalignment look flat in the rigid-body frequency range. If we double the mass, and the rigid-body acceleration under the same force reduces two times, the inertia force, and therefore relative motion and misalignment under the force excitation stay the same. In the case of kinematic excitation from the foundation (Figs. 7.17 and 7.18), the rigid-body transmissibility stays the same, so doubling the mass (and therefore the inertia forces) leads to doubling of the relative displacement and misalignment.

The effect of increasing bending stiffness, as well as simultaneous increasing of mass and stiffness (adding structural mass) can be researched in the same manner. Figures 7.21–7.25 illustrate the transmissibilities, dynamic compliances, and misalignments under kinematic and force excitations for different scenarios of increasing stiffness or simultaneous increase of mass and stiffness. (Relative flexural displacements behave generally the same way as misalignments,

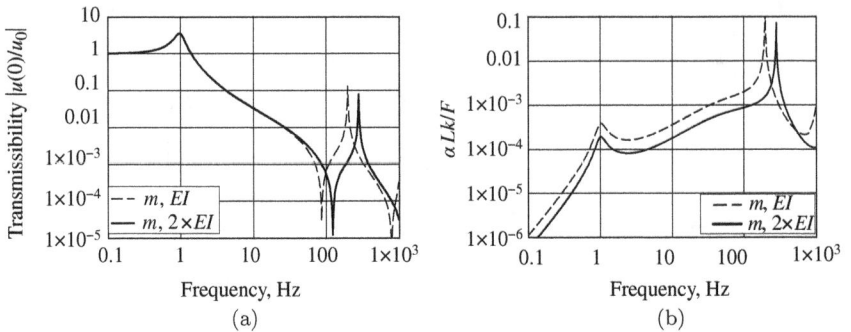

**Figure 7.21.** Example 7.1: Positive effect of increased bending stiffness (a) on the vibration transmissibility at the edge of the beam, $u(0)/u_0$, and (b) on the maximum misalignment, $\alpha L$, of the pneumatically isolated platform under the kinematic excitation $u_0$ from the floor (Fig. 7.11(a)). Rigid-body frequency range expands; resonance amplitude decreases; quasi-static deformation decreases.

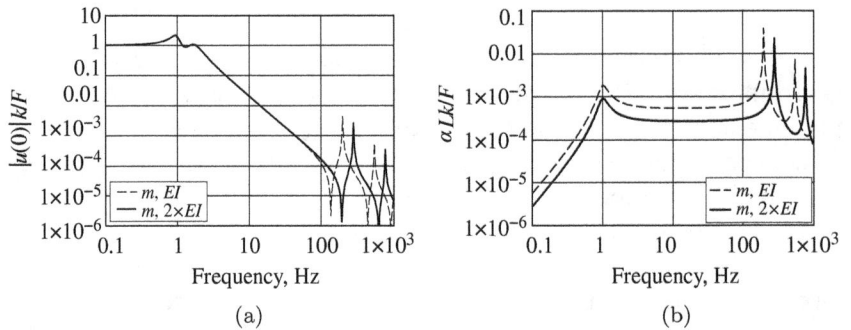

**Figure 7.22.** Example 7.1: Positive effect of increased bending stiffness on the dynamic compliance (a) and the maximum misalignment (b) of the pneumatically isolated beam under the force excitation $F$. Both functions are reduced to non-dimensional form by referring to $1/k$. Rigid-body frequency range expands; rigid-body line stays the same; resonance frequencies increase; resonance amplitudes decrease; quasi-static deformation decreases.

so we do not show these plots for sake of brevity.) It is necessary to increase the bending stiffness at least in the same proportion as mass (in this case, two times) to avoid the negative effects of increasing inertia forces under kinematic excitation. In this case transmissibility and relative motion under kinematic excitation from the floor stay the same as shown in Figs. 7.17 and 7.18, hence no new graph is given for this case; however, the dynamic response under force excitation does improve (see Fig. 7.23). Increasing bending stiffness

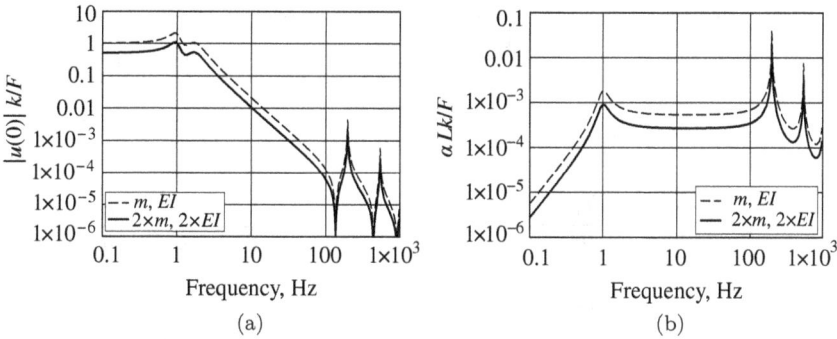

**Figure 7.23.** Example 7.1: Effect of structural mass (bending stiffness and mass increased in the same proportion): (a) on the dynamic compliance, $u(0)/F$, and (b) on the maximum misalignment $\alpha L$ of the pneumatically isolated beam under the force excitation $F$. Both functions are reduced to non-dimensional form by referring to $1/k$. Rigid-body line shifts down; resonance frequencies stay the same; resonance amplitudes decrease; quasi-static bending decreases.

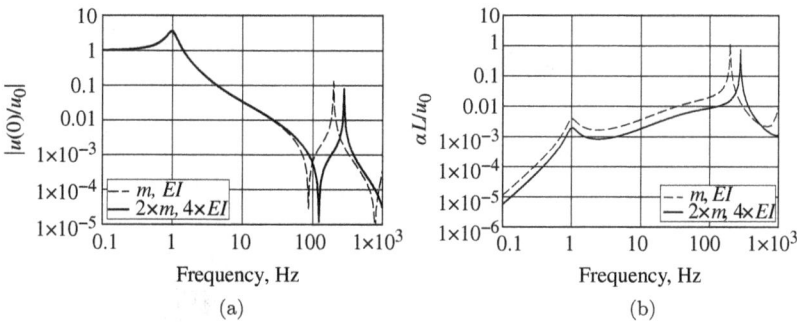

**Figure 7.24.** Example 7.1: Positive effect of bending stiffness increased in higher proportion than the mass: (a) on the vibration transmissibility at the edge of the beam, $u(0)/u_0$, (b) on the maximum misalignment $\alpha L$ under the kinematic excitation $u_0$ from the floor. Rigid-body frequency range expands; resonance amplitude decreases; relative motion decreases.

invariably leads to improvement in dynamic qualities; the best results are obtained when stiffness increases simultaneously with mass but in higher proportion (Figs. 7.24 and 7.25 show the result of increasing the mass two times and the bending stiffness four times).

In application to real-life isolated platforms, the latter case of Example 7.1 comes close to describing the effect of increasing the thickness of the optical table. The effect of thickness on the table performance is illustrated by Fig. 7.26 that depicts experimental

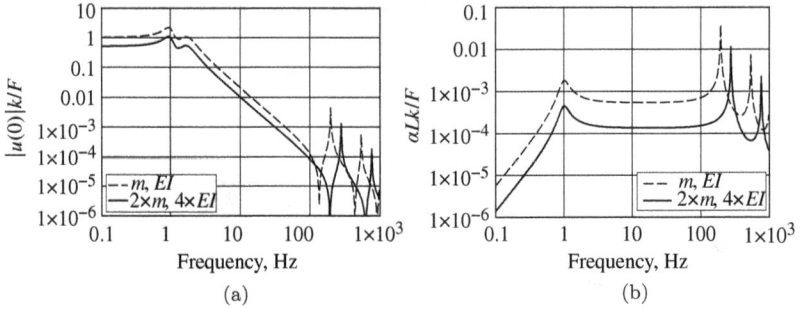

**Figure 7.25.** Example 7.1: Positive effect of bending stiffness increased in higher proportion than the mass on the dynamic compliance $u(0)/F$ (a) and the misalignment per unit force $\alpha L/F$ (b) of the pneumatically isolated platform under the force excitation (Fig. 7.11(b)). Both functions are reduced to non-dimensional form by referring to the static compliance in the base case, $1/k$. Rigid-body frequency range expands; rigid-body line moves down; resonance amplitudes decrease; quasi-static deformation decreases.

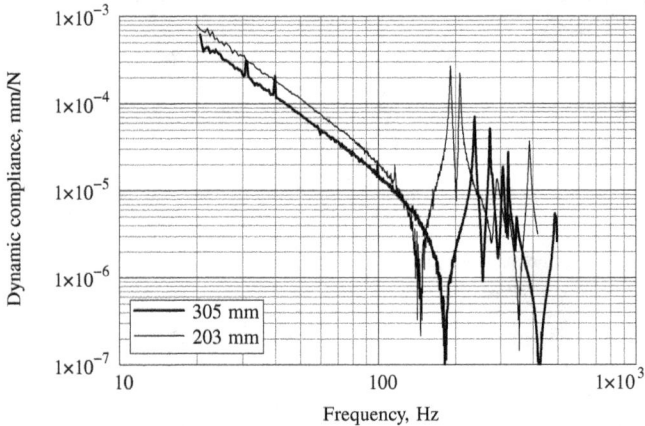

**Figure 7.26.** Experimental corner dynamic compliances of optical tables 1.22 m × 2.44 m with different thicknesses. The tables did not use any additional means of damping except structural damping.

dynamic compliances of the 305 mm thick and 203 mm thick optical tables of the same width and length and the same design. The thicker table has significantly higher resonance frequencies and lower resonance amplitudes.

Although this model example captures the main physical properties of pneumatically isolated platforms, there is a substantial difference from real optical tables. The most significant difference is that a table, being a plate-like structure able to deform in three dimensions, has a much denser spectrum of resonance frequencies. Several resonances can be usually observed in the 100–1000 Hz range.

*Figures of merit.* Dynamic compliance curves are generally accepted as indicators of dynamic quality of optical tables. Attempts were made to reduce the information provided by the compliance curves to a single numerical criterion, or a "figure of merit", that would allow for comparison of dynamic qualities of various isolated platforms and estimating misalignments under typical external conditions. Accurate prediction of flexural displacements of the table due to the floor vibration would require detailed analysis considering individual vibration inputs at each leg, vibration attenuation through legs, dynamic compliances at the support points, and transfer functions from support points to individual points of the table. All this information is not readily available.

Newport Corporation (now a subsidiary of MKS) recommends [12] a simplified approach extending the formulas for root mean square (RMS) displacements derived in Chapter 3 (see Eqs. (3.26), (3.38) and (3.60)) to estimate the RMS relative motion of two remote points on the surface of the optical table due to resonance response at a certain natural frequency $f_n$. The RMS formula is augmented to include the effect of vibration isolation (by employing the transmissibility $T$) and the effect of relative motion (by adding the factor of 2). This leads to the following estimate for the maximum relative motion (MRM) of two points on the surface of the table:

$$\text{MRM}_n = \sqrt{\frac{1}{32\pi^3}} \sqrt{\frac{Q}{f_n^3}} \sqrt{\text{PSD}} \cdot g \cdot T \cdot 2. \tag{7.15}$$

Here, the resonance amplification $Q$ (equal to $1/\eta$ for a single-degree-of-freedom model) depends on loss factor and modal participation; for practical purposes, for a few first flexural resonances it can be estimated by the ratio of resonance amplitude to the "rigid body" amplitude measured from the corner compliance as shown in Fig. 7.27. (Figure 7.27 is a blown-up and marked version of Fig. 7.9.) PSD means power spectral density of acceleration; $g$ is the acceleration of gravity, $g = 9.8$ m/s$^2$. The numerical value of PSD in $g^2$/Hz is used in (7.15). The values of PSD listed in Table 5.3 of Chapter 5 are recommended as typical.

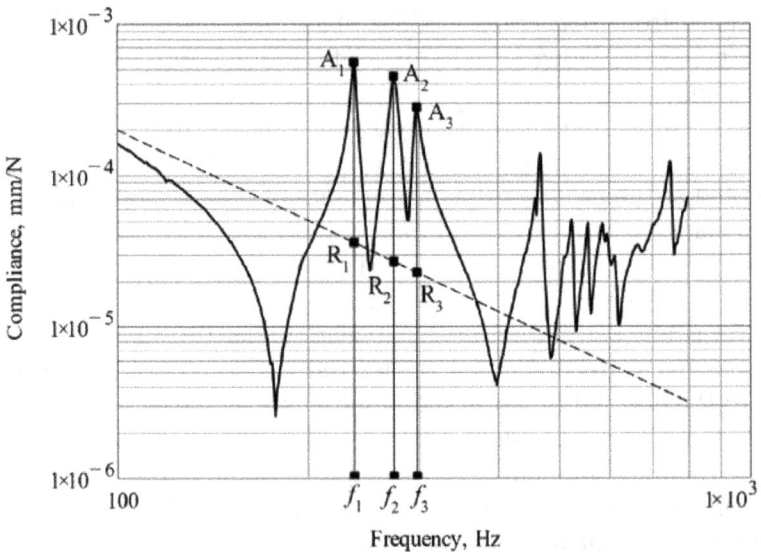

**Figure 7.27.** Illustration to Eq. (7.15) and Exercise 7.5. $Q_n = A_n/R_n$.

It is recommended to use, as a typical value of transmissibility, $T = 0.01$. The result is relative motion in meters. Equation (7.15) combines all factors influencing the dynamic reaction of the table: intensity of floor vibration, quality of vibration isolation, resonance frequency, and damping. The value

$$\text{DDC}_n = \sqrt{\frac{Q}{f_n^3}}, \tag{7.16}$$

that depends on the dynamic quality of the platform alone, is called a Dynamic Deflection Coefficient.

For highly damped structures the base assumption of RMS estimates ($\eta \ll 1$) is not valid; still the criteria (7.15) and (7.16) can be used as a "figure of merit," except it is recommended to replace $Q$ by $\sqrt{Q^2 - 1}$.

**Exercise 7.5.** Estimate the relative motion of the table using Eq. (7.15) and experimental dynamic compliance from Fig. 7.27 assuming typical average acceleration level on the floor PSD $= 10^{-8} g^2/\text{Hz}$.

**Solution.** We'll estimate contributions of three main resonances. From this graph, the resonance frequencies are $f_1 = 237\,\text{Hz}$,

$f_2 = 274$ Hz, $f_3 = 298$ Hz, with the corresponding spectral amplitudes $A_1 = 5.55 \cdot 10^{-4}$ mm/N, $A_2 = 4.49 \cdot 10^{-4}$ mm/N, and $A_3 = 2.80 \cdot 10^{-4}$ mm/N. The "rigid body" compliances, at the same frequencies, are $R_1 = 3.56 \cdot 10^{-5}$ mm/N, $R_2 = 2.66 \cdot 10^{-5}$ mm/N, and $R_3 = 2.25 \cdot 10^{-5}$ mm/N. This gives $Q_1 = A_1/R_1 = 15.6$; $Q_2 = A_2/R_2 = 16.8$; $Q_3 = A_3/R_3 = 12.5$. Dynamic Deflection Coefficients, according to (7.16), are:

$$\text{DDC}_1 = \sqrt{\frac{Q_1}{f_1^3}} = 1.1 \cdot 10^{-3}, \quad \text{DDC}_2 = \sqrt{\frac{Q_2}{f_2^3}} = 0.9 \cdot 10^{-3},$$

$$\text{DDC}_3 = \sqrt{\frac{Q_3}{f_3^3}} = 0.7 \cdot 10^{-3}.$$

MRM values are calculated by the formula (7.15) as follows, in meters:

$$\text{MRM}_1 = \sqrt{\frac{1}{32\pi^3}} \sqrt{\frac{15.6}{237^3}} \sqrt{10^{-8}} \cdot 9.8 \cdot 0.01 \cdot 2 = 6.7 \cdot 10^{-10},$$

$$\text{MRM}_2 = \sqrt{\frac{1}{32\pi^3}} \sqrt{\frac{16.8}{274^3}} \sqrt{10^{-8}} \cdot 9.8 \cdot 0.01 \cdot 2 = 5.6 \cdot 10^{-10},$$

$$\text{MRM}_3 = \sqrt{\frac{1}{32\pi^3}} \sqrt{\frac{12.5}{298^3}} \sqrt{10^{-8}} \cdot 9.8 \cdot 0.01 \cdot 2 = 4.3 \cdot 10^{-10}.$$

This estimate suggests that the contributions of the first three resonances to the motion of the table are of the same order of magnitude. Contributions of higher resonances are significantly smaller. The total relative motion can be estimated, assuming independent inputs in the three frequency ranges, by the Pythagorean sum of the three resonance responses: $\text{MRM} = \sqrt{\text{MRM}_1^2 + \text{MRM}_2^2 + \text{MRM}_3^2}$, which gives, approximately, 1 nanometer.

Dynamic vibration absorbers, or tuned mass dampers (TMD), whose principle of action is described in Section 7.2, find wide applications in damping internal resonances of vibration-isolated platforms.

**Example 7.2. Dynamic compliance and transmissibility of an isolated platform supplied with TMDs.** This example was originally developed as part of Newport Corporation's technical note. The simplified model of an isolated platform considered in Example 7.1

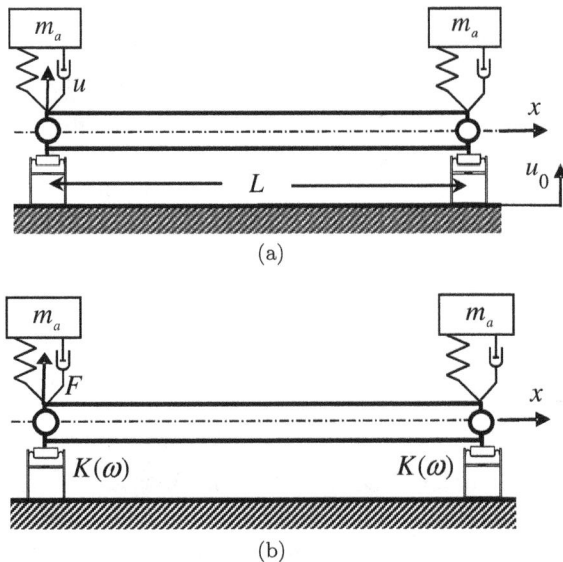

Figure 7.28. Example 7.2: Model of an isolated platform with dynamic vibration absorbers (TMD). (a) Kinematic excitation; (b) force excitation.

allows for simulation of the effect of dynamic vibration absorbers, or tuned mass dampers (TMD). They can be modeled as damped oscillators attached to the ends of the beam (Fig. 7.28).

The dynamic compliance of the damped oscillator at the attachment point was derived in Section 3.2 of Chapter 3. For this example, we'll need the expression for dynamic stiffness, that is, the ratio of complex amplitudes of force acting on the oscillator at the attachment point to the displacement at this point:

$$K_a(\omega) = \frac{-m_a\omega^2}{1 - \frac{\omega^2}{\omega_a^2(1+i\eta_a)}},$$

where $m_a$ is the mass of the damper, $\omega_a$ is the partial frequency of the damper (tuning frequency), $\eta_a$ is the loss factor. The frequency-dependent boundary conditions for the beam equation (7.11) are given by (7.12) and

$$EIu'''(0) = -K(\omega)(u(0) - u_0) - K_a(\omega)u_0;$$
$$EIu'''(L) = K(\omega)(u(L) - u_0) + K_a(\omega)u_0$$

in case of kinematic excitation of the foundation with the amplitude $u_0$, Fig. 7.28(a), or

$$EIu'''(0) = -[K(\omega)+K_a(\omega)]u(0)+F; EIu'''(L) = [K(\omega)+K_a(\omega)]u(L)$$

in case of force excitation, Fig. 7.28(b). Further calculations of the complex amplitudes $u(x)$ are straightforward, following the procedure outlined in Example 3.5 of Chapter 3.

Parameters of the dynamic absorbers may be determined based on their mass ratios (7.9) according to the optimization formulas (7.5), (7.6). Assume for the moving masses of each absorber $m_a = 0.025\ m$ (2.5% of the mass of the platform). Direct calculation gives effective modal masses of the beam $m_k = 0.25\ m$ for all modes. Figures 7.29 and 7.30 illustrate transmissibilities and dynamic compliances for two cases: first, both dampers are tuned to the first resonance; second, one damper (at $x = 0$) tuned to the first resonance and the second damper (at $x = L$) — to the second resonance. In the first case, mass ratio is $\mu_k = 2m_d/m_k = 0.2$. Therefore, the dampers tuned to the first flexural resonance, $f_1 = 200$ Hz, should have, according to (7.5), partial resonance frequency 167 Hz; the optimal loss factor is $\eta = 0.56$ according to (7.6). In the second case $\mu_k = m_d/m_k = 0.1$ for both resonances, so, the damper tuned to the first flexural resonance, $f_1 = 200$ Hz, should have, according to (7.5), partial resonance frequency 182 Hz; the damper tuned to the second flexural resonance, $f_2 = 551$ Hz, partial resonance frequency 501 Hz. The optimal loss factor is $\eta = 0.39$ according to (7.6). As in Example 7.1, we assume that

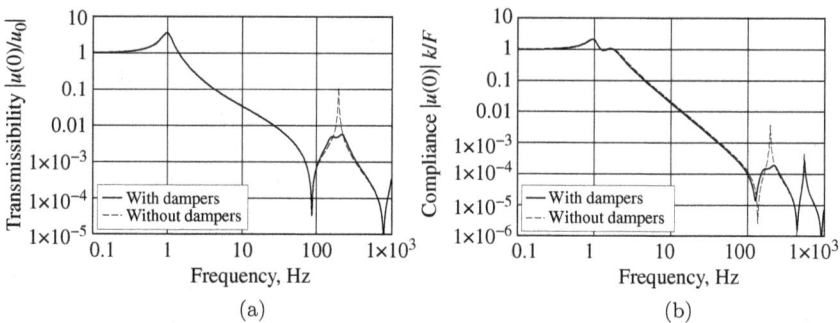

**Figure 7.29.** Example 7.2: Transmissibility (a) and dynamic compliance (b) of the model of an isolated platform (Fig. 7.28) with both dampers tuned to suppress the first flexural resonance.

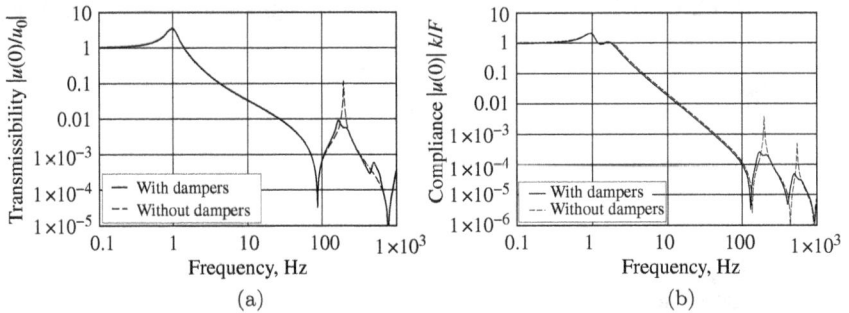

**Figure 7.30.** Example 7.2: Transmissibility (a) and dynamic compliance (b) of the model of an isolated platform (Fig. 7.28) with one damper tuned to suppress the first flexural resonance, another one — the second flexural resonance.

the stiffness of pneumatic isolators adjusts to a slightly increased weight to keep the "bouncing" frequency constant. The graphs show that in case when both dampers are tuned to the first resonance peak, it is reduced almost 20 times, however, the second peak of the dynamic compliance stays almost the same. When dampers are tuned to separate peaks, both resonances are suppressed, but to a lesser extent. Peak values of relative displacement and misalignment are suppressed by dynamic vibration absorbers accordingly.

This model example shows that tuned mass dampers can greatly reduce deviations from rigid-body behavior. The technology finds wide applications to vibration-isolated platforms for optical setups.

Tuned mass dampers designed specifically for optical tables, and examples of their applications, are described in [11, 12]. Figure 7.31 shows a contemporary design of a tuned mass damper for an optical table. In this design, a steel block, serving as the moving mass of the damper, is supported symmetrically along the two ribs at the top and two ribs at the bottom by steel plates acting as leaf springs. The springs are pressed against profiled loading plates with pre-load exerted by a loading screw. As the plate bends due to pre-load, the line of contact with the profiled support shifts, so the effective length of the leaf spring changes, which, in turn, changes the stiffness and the natural frequency of the dynamic absorber. Thin strips of highly damped elastomeric material are bonded to the surface of each plate in the contact areas. The elastomer forms a thin damping layer between the spring and the loading plate that does not affect

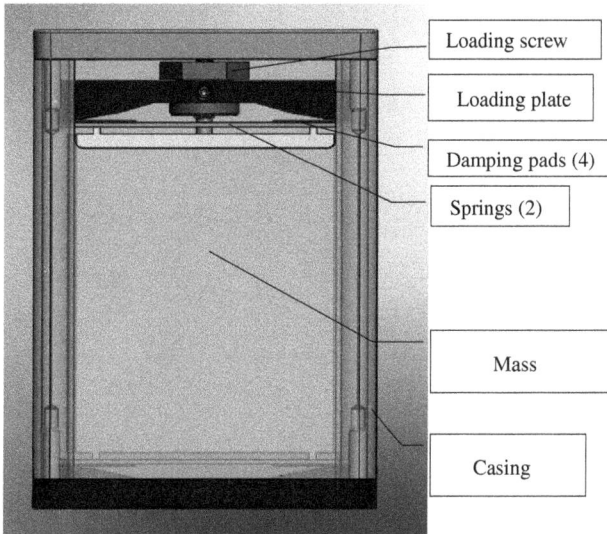

**Figure 7.31.** Adjustable tuned mass damper for optical tables [11].

significantly the stiffness (due to near incompressibility of the elastomeric material working in compression) but ensures the necessary dissipation of energy due to shear deformation across the layer. As a result, the stiffness and damping functions are split between the leaf spring and the damping layer, thereby ensuring thermal stability of stiffness, as well as adequate damping.

With a proper radius of the profiled supports, the stiffening-up of the springs leads to almost linear frequency vs. pre-load dependence. Practically, however, what matters is that it is possible to cover the desired range of frequencies continuously by operating the pre-loading mechanism. This is illustrated in Fig. 7.32. This damper can be tuned to a required undamped frequency using only one sensor without access to the moving mass [11].

Damping of resonance response of an optical table by TMDs is illustrated in Fig. 7.33. It is not unusual to have the peak amplitude resonance amplitude reduced ten times or more. Other practical applications of TMDs and design examples can be found in the monograph [2].

A *breadboard* is a smaller optical platform, usually no longer than 1.5 m and no thicker than 110 mm. Breadboards are used increasingly often for vibration-sensitive optical applications as small footprint,

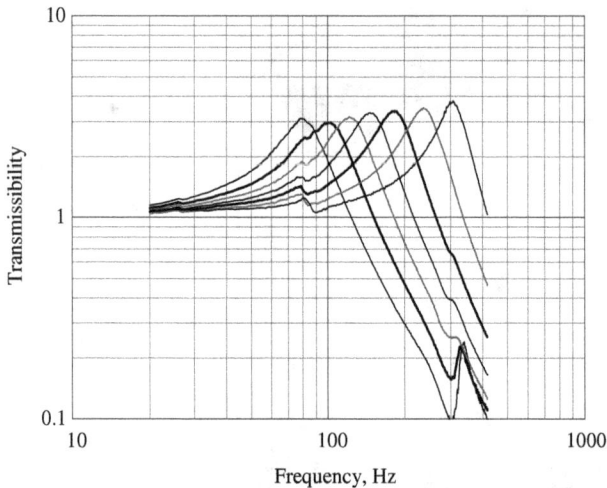

**Figure 7.32.** A family of experimental transmissibilities describing the tuning of the TMD. Smooth tuning of the resonance frequency by operating the loading screw covers a range of frequencies (80–300 Hz) with a small variation of the loss factor [11].

low-cost alternatives to larger optical tables. Applying the TMD technology to thinner platforms presents design challenges due to lack of space between the facesheets for relatively massive absorbers. Another method of introducing damping, more suitable for smaller breadboards, is described below following the article [13]. It makes use of shear deformation in a thin layer of highly damped viscoelastic material in conjunction with specific modal geometry of honeycomb platforms.

High damping occurs when there is shear deformation across a highly dissipative material such as a viscoelastic elastomer. Dynamic finite-element analysis of a honeycomb platform reveals areas of maximum shear deformation. Figure 7.34 shows the geometrical patterns of first (twisting) and second (bending) modes of flexural vibration. Areas of higher shear deformation are shown in darker tones. However, simply filling these areas with a highly damped compound would not be effective enough. The dissipated energy would be insignificant compared to the elastic energy stored in the elastically deformed facesheets. The practice of vibration damping teaches that the deformation across viscous elements must be amplified in

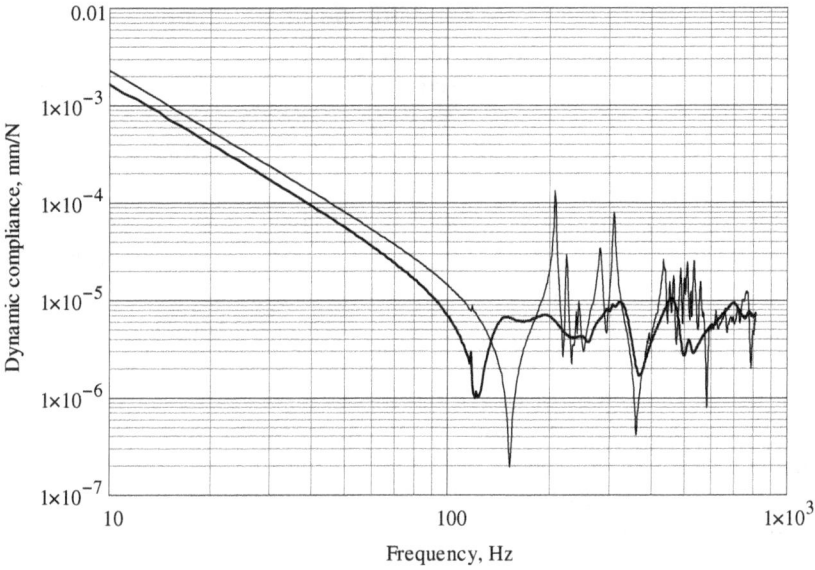

**Figure 7.33.** Application of tuned mass dampers to an optical table. Corner compliance is shown for a table 1.52 m × 2.44 m × 305 mm with six dampers integrated into the table structure at the corners. Thin line — without dampers, thick line — with dampers.

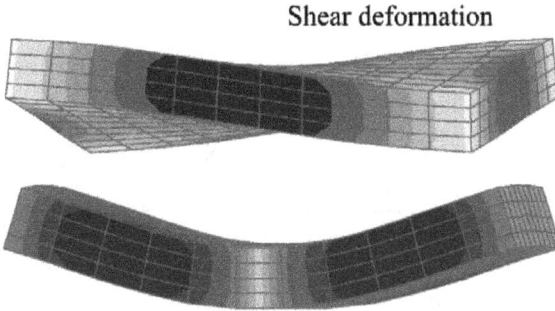

**Figure 7.34.** Finite-element representation of two first resonance modes of an optical breadboard. Dark color shows areas of maximum shear deformation [13].

order to dissipate enough energy. In case of a dynamic absorber (Section 7.2), for example, it is done dynamically, using the resonance of the auxiliary mass; in case of constrained layer damping (Fig. 7.2) it is done kinematically by restricting the tensile strain in the viscoelastic layer. In the present case, specific geometry of the normal modes is used. As seen in Fig. 7.34, the deformation of the breadboard does not conform to the classic Kirchhoff theory of plates requiring that the normals to the middle surface remain straight and perpendicular; rather, top and bottom facesheets in the dark highlighted areas rotate in unison about parallel horizontal axes. That prompts the damping arrangement shown in Fig. 7.35. One metal plate is attached to the top facesheet, another to the bottom facesheet, and a layer of highly damped elastomer is sandwiched between the two plates. When the plates turn, this small rotation translates into substantial mutual shift, and therefore into large shear deformation across the damping layer (Fig. 7.36). Several of those damping assemblies can be placed around the periphery of the breadboard.

The main design parameters of the damping assembly are the shear modulus of elastomer, $G$, the shear modulus of metal, $G_m$, the thickness of elastomer, $h$, the thickness of metal, $h_m$, and the loss factor of elastomer, $\eta$. The design goal is to maximize the energy $E$ dissipated per cycle:

$$E = 2\pi\eta \cdot \frac{1}{2} G H_1 L h \gamma^2.$$

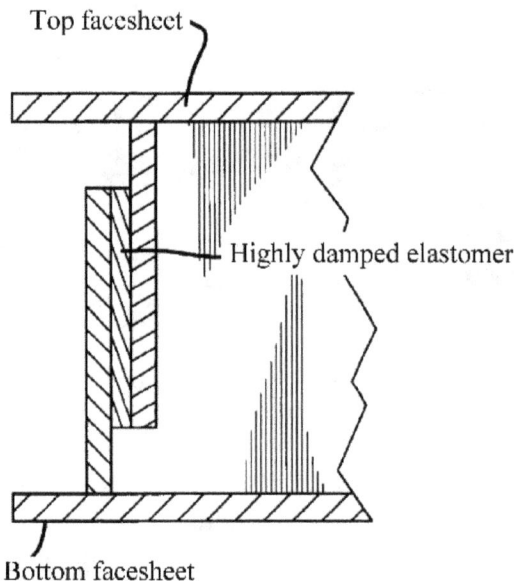

Top facesheet

Highly damped elastomer

Bottom facesheet

**Figure 7.35.** Schematic of the damping treatment. Adapted from [14].

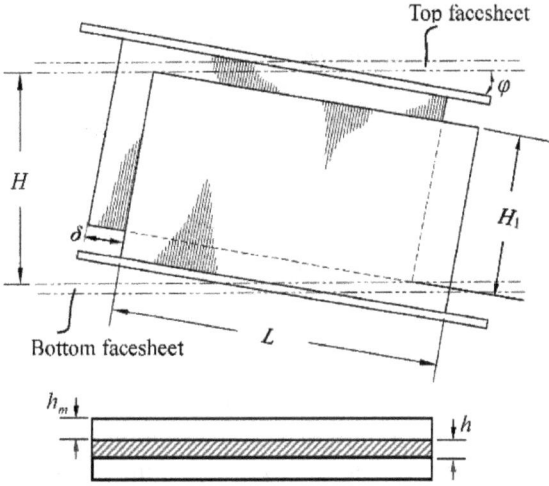

**Figure 7.36.** Mechanism of shear deformation. Adapted from [14].

Here $\gamma$ is the shear deformation across the elastomer,

$$\gamma = \frac{H}{h}\varphi_e,$$

where $\varphi_e$ is the effective rotation angle (the rotation angle of the faceplates minus the skew angle of metal plates due to their elastic deformation). In terms of the parameters of the elastomer, maximizing $E$ requires increasing the modulus $G$ and decreasing the thickness $h$; on the other hand, the design must ensure that the deformation involves mostly shear across the damping material, not the in-plane deformation of the metal plates. If the layer is too thin and too stiff, the assembly would be reduced to two plates glued together. So, design optimization is necessary. Assuming a simplified deformation model, namely pure shear in all elements, one can express the dissipated energy in term of the rotation angle, $\varphi$, by the following equation:

$$E = \pi\eta LHG_m h_m \frac{\alpha}{(1+2\alpha)^2}\varphi^2,$$

where

$$\alpha = \frac{G}{G_m}\frac{HH_1}{hh_m}.$$

Maximizing $E$ leads to $\alpha = 0.5$. This simplified analysis allows one to verify that realistic values of parameters can lead to the damping close to optimal. In real life we usually do not get to optimize the design: we can only choose from several design options tied to highly damped elastomers commercially available in form of thin sheets, and standard gages of metal plates. Note that increasing thickness and weight of the restraining metal plates is not desirable because it would lead to lower resonance frequencies. An effective tool for analyzing multiple design options is provided by finite element analysis. In particular, NASTRAN offers solution sequence 101 "Complex Eigenvalues" that allows for fast calculation of natural frequencies and corresponding loss factors. The model does not need to be very detailed since the goal of the analysis is not a precise description of normal modes but rather a comparison of natural frequencies and modal loss factors between different options. Since the normal modes can change when the breadboard carries a load, the laminated damped plates are placed along all four sides of the breadboard. They play the cosmetic role of edge finish. Specific design based on this principle is given below in Example 7.3.

**Example 7.3.** This example considers the results of the development of the damped optical breadboard according to [14] for the 0.9 m × 1.5 m × 110 mm breadboard. Without going into technical details of the finite-element modeling and analysis, the calculated modal properties of the design identified after analyzing multiple cases are shown in Table 7.1. This combination of parameters corresponds to $\alpha = 0.41$.

According to the analysis, significant loss factors can be achieved for all modes. Figure 7.37(a) shows the sketch of the final design of the damped edge finish. Two layers of elastomer (see Fig. 7.37(b)) are incorporated in 59 mm thick breadboards using the same principle.

Figure 7.38 compares experimental dynamic compliances of two vibration-isolated breadboards of the same size, one with standard edge finish, another with laminated edge finish described above.

**Table 7.1.**   Modal properties of the design model.

| Frequency (Hz) | 264.3 | 294.7 | 523.4 | 602.2 | 642.8 | 715.1 |
|---|---|---|---|---|---|---|
| Modal loss factor | 0.087 | 0.014 | 0.106 | 0.015 | 0.044 | 0.038 |

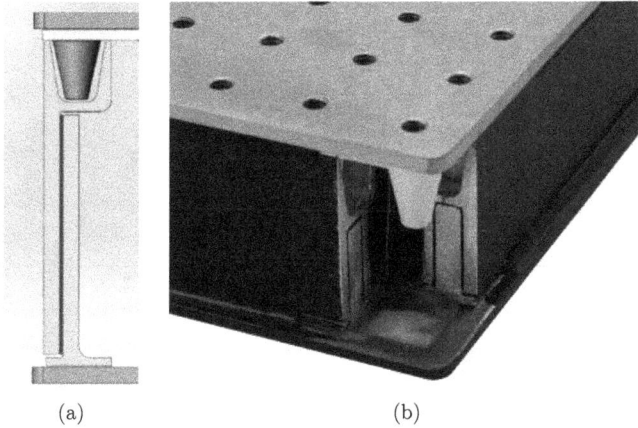

(a)               (b)

**Figure 7.37.** Example 7.3: (a) Sketch design of the damped edge finish for the 110 mm breadboard (accommodations are made for spill-proof sealing cups); (b) Photo of a 59 mm breadboard with damped edge finish (corner piece removed for sake of illustration). Courtesy of Newport Corporation. All rights reserved.

The comparison shows that introducing the laminated edge finish reduces all resonance peaks significantly, the main one about nine times. The observed damping factors are in general agreement with the predictions of Table 7.1. The reduction in resonance response leads to significantly improved protection of vibration-sensitive optical payloads.

This damping solution presents a specific form of a damping link [1], where one creates damping by connecting, using a vibration-absorbing element, two areas of the structure that experience relative motion. In this case, angular motion is transformed by the link into shear deformation. The method can be called "modal damping" since it uses the particular geometry of mode shapes. It is important to note that the same method of damping can be applied to other structures such as frames and trusses.

## 7.4 Terminological Note

In conclusion of the last two chapters, a terminological remark may be in order. There is some ambiguity in technical communications concerning the terms "vibration isolation" and "vibration damping". Oftentimes, people refer to any reduction of vibration as damping.

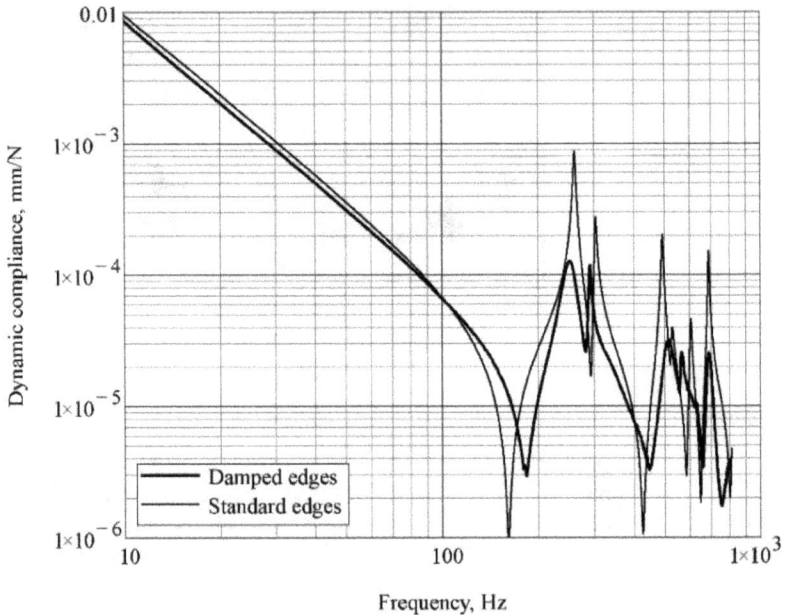

**Figure 7.38.** Dynamic corner compliance of the breadboard with damping treatment by laminated edges and the breadboard with standard composite edges of the same size.

Vibration damping means dissipation of mechanical energy into heat. Vibration isolation means reflecting mechanical energy back to the source, usually by a barrier created from low-stiffness links (isolators). These are two different physical phenomena and two different methods of vibration control.

There can be isolation in absence of damping: a simple undamped elastic support provides isolation at frequencies higher than natural frequency times $\sqrt{2}$ (see Chapter 3). On the other hand, a non-isolated (rigidly supported) flexible structure will exhibit a lower amplitude of resonance vibration if it incorporates some mechanisms of damping. That being said, a properly designed vibration isolator must include some means of energy dissipation to avoid unlimited resonance amplification. There may be a trade-off between isolation and damping. Figures 3.6 and 3.11 of Chapter 3, as well as Fig. 6.22 of Chapter 6, show examples of such trade-offs; see also Section 8.5 of Chapter 8.

Speaking about vibration-isolated platforms, we distinguish between two main components: soft isolators (legs) and a stiff tabletop. The isolators (with their own incorporated means of damping) are responsible for preventing the floor vibration from reaching the tabletop. Even if the isolators work ideally, the tabletop may be susceptible to resonance vibration caused by residuals from the floor vibration or acoustical and on-board excitation, and therefore damping should be introduced there in form of integrated damping or tuned mass dampers.

Design of a vibration isolator and design of a vibration damper are two very different tasks, although both may share the same elements.

# References

1. D.I.G. Jones, Applied damping treatments, in C.M. Harris (ed.), *Shock and Vibration Handbook*, 4th edn., McGraw-Hill, New York, 1995, pp. 37.5–37.10, and subsequent editions.
2. A.D. Nashif, D.I.G. Jones, and J.P. Henderson, *Vibration Damping*, Wiley, New York, 1985.
3. R. Ross, E.E. Ungar, and E.M. Kerwin, Damping of plate flexural vibration by means of viscoelastic laminae, *Structural Damping*, ASME Publication, New York, 1959, pp. 49–88.
4. F.E. Reed, Dynamic vibration absorbers and auxiliary mass dampers, in C.M. Harris (ed.), *Shock and Vibration Handbook*, 4th edn., McGraw-Hill, New York, 1995, pp. 6.7–6.16, and subsequent editions.
5. B.G. Korenev and L.M. Reznikov, *Dynamic Absorbers of Vibrations. Theory and Technical Applications.* Nauka, Moscow, 1988, 303 p. (in Russian). English translation: Wiley, New York, 1993, 312 p.
6. T. Asami, O. Nishihara, and A.M. Baz, Analytical solutions to $H_\infty$ and $H_2$ optimization of dynamic vibration absorbers attached to damped linear systems, *Journal of Vibration and Acoustics*, vol. 124, pp. 284–295, 2002.
7. O. Nishihara and T. Asami, Closed-form solutions to the exact optimizations of dynamic vibration absorbers (minimizations of the maximum amplitude magnification factors), *Journal of Vibration and Acoustics*, vol. 124, pp. 576–582, 2002.
8. M. Zilletti, S.J. Elliott, and E. Rustighi, Optimisation of dynamic vibration absorbers to minimise kinetic energy and maximise internal power dissipation, *Journal of Sound and Vibration*, vol. 331, pp. 4093–4100, 2012.

9. S. Krenk and J. Høgsberg, Tuned mass absorber on a flexible structure, *Journal of Sound and Vibration*, vol. 333, pp. 1577–1595, 2014.

10. S. Krenk and J. Høgsberg, Tuned resonant mass or inerter-based absorbers: unified calibration with quasi-dynamic flexibility and inertia correction, *Proc. Royal Society A*, vol. 472, 2016; http://dx.doi.org/10.1098/rspa.2015.0718.

11. V.M. Ryaboy, Practical aspects of design, tuning and application of dynamic vibration absorbers, *Proceedings of Meetings on Acoustics*, vol. 26, 065006, 2016; https://doi.org/10.1121/2.0000231.

12. *Newport Resource*, Catalog, Newport Corporation, 2004 and later editions.

13. V.M. Ryaboy, Damping treatment of optical breadboards, *Proceedings of Meetings on Acoustics*, vol. 30, 065010, 2017; https://doi.org/10.1121/2.0000581.

14. Vibration damper for optical tables and other structures. US Patent 6,598,545, 2003.

Chapter 8

# System Approach to Vibration Control: Optomechanical Case Studies and Troubleshooting

System approach implies paying attention to the entire range of causes and effects important for resolving the problem at hand. It is hard to overestimate the importance of system approach in vibration control. Sometimes, when higher than acceptable vibration is noticed, attempts to remedy the situation are concentrated exclusively on the available vibration control devices. This is a mistake because it lacks the system approach. In this chapter, we analyze the role of different sub-systems in vibration sensitivity of optical applications and learn the practice of vibration troubleshooting through case studies.

## 8.1 Role of Different Sub-systems in Vibration Sensitivity of Optical Applications

Dynamic properties of the whole structural path that includes a support structure (table), optomechanical elements, on-board instrumentation, optics, and motion control systems are important for the stability of an optical application.

Typically, an optical table or other isolated platform serves as a common base for the whole optomechanical assembly. Optomechanical components including posts, rods, lens holders, and mirror mounts, as well as positioning stages, must anchor the optical elements so that the optical paths are undisturbed by environmental impacts such as vibration. The resulting performance depends on the overall "structural loop" that encompasses support structures, motion control systems, and optomechanical elements. Each element has its own dynamic response; they interact through their structural connections characterized by (often complicated) transfer functions.

Various sources of vibration and weak points in vibration transmission paths can be identified by meticulous testing. The following case studies elucidate various aspects of dynamic behavior of optomechanics and ways to troubleshoot and improve the dynamic qualities of optomechanical assemblies.

## 8.2  Case Study: Table Resonances and Optomechanical Resonances

Dynamic deflection of optomechanical components can be a substantial contributor to unwanted vibration in optical systems. Even in environments where the optical table is acting as a rigid body, this rigid-body motion can excite vibrations of optical mounts and cause beam instability. Our first case study considers an experiment intended to analyze the sensitivity of an optomechanical element to vibration. We examine a 12.7 mm (0.5 in.) diameter optical post mounted inside a post holder so that the total height equals 152.4 mm (6 in.). An accelerometer is attached to the top of the post through an aluminum cube to measure the motion in the horizontal direction (Fig. 8.1). The total mass of the cube and the accelerometer is close to the typical mass of an optical mount.

A dynamic force was applied to the tabletop by a miniature electrodynamic shaker through a force sensor. The dynamic performance of the post holder – post assembly was characterized by the transfer dynamic compliance, which is the ratio of the displacement measured by the accelerometer in the horizontal direction at the top of the post, to the excitation force applied to the surface of the table, as a function of frequency. It shows the natural frequencies and the level of damping of the assembly. The test was performed on the surface of an optical table called SmartTable® [1]. Details of design

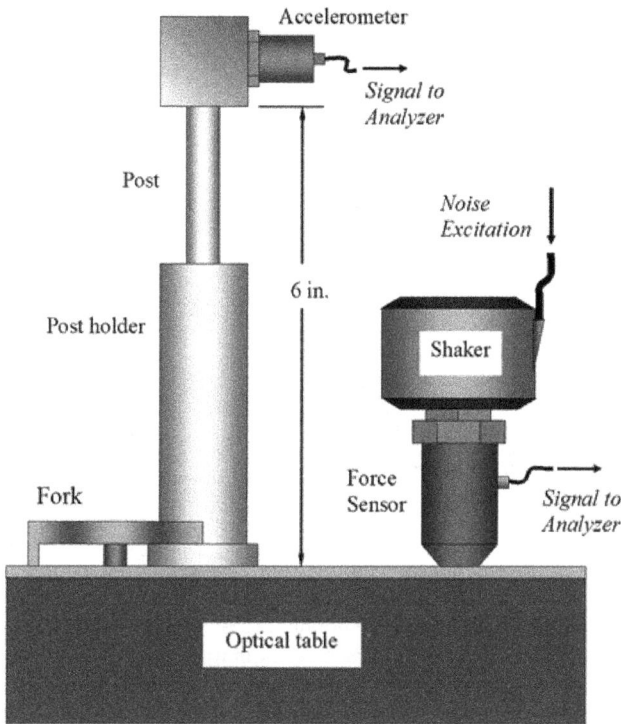

**Figure 8.1.** Experiment illustrating the relative importance of resonance responses of optomechanics and the platform [2].

and performance of the SmartTable are given in the next chapter; for the purpose of this section note that it includes electronic sensing and actuation, therefore, the dissipation of mechanical energy can be switched on and off in the same physical structure. That makes it, besides direct applications, a valuable learning tool about the role of damping.

This setup allows one to appreciate the effect of structural damping in the table and distinguish between the resonances of the table and that of the post–post holder system. Analysis of the measured vibration transfer functions with active damping enabled and disabled (Fig. 8.2) clearly shows two resonance peaks generated by local vibration modes of the post–post holder system and one prominent resonance that is due to the table itself. Though active damping of the table suppresses global vibration modes of the tabletop, the local

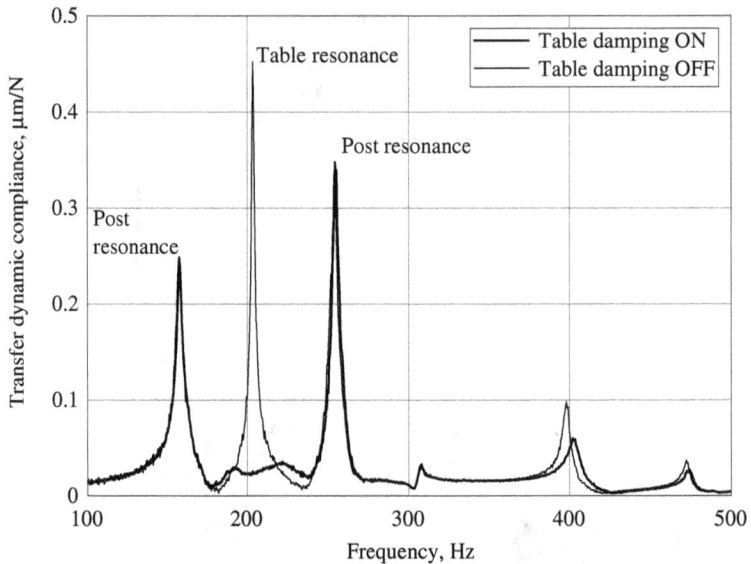

**Figure 8.2.** Experimental results illustrating the relative importance of table resonances and optomechanical component resonances (test setup shown in Fig. 8.1). Even if the table resonances are eliminated, optomechanical resonances can be excited by environmental impacts. Adapted from [2].

modes of the post–post holder system are still present. The experiment shows that the dynamics of optomechanical elements can be as important as the dynamics of the table.

Several approaches can be used to control this contribution: reduce optical axis heights whenever possible, use the most robust posts and clamps available for sensitive optics, and place these sensitive components as close as possible to the center of the table where the dynamic compliance is typically the lowest. If it is necessary to arrange a higher optical axis, large diameter damped optical posts or rods should be used as explained in the next sections.

## 8.3 Case Study: Geometry and Dynamics of an Optical Post

The case study described in the previous section shows how important is the dynamics of optomechanical elements such as optical

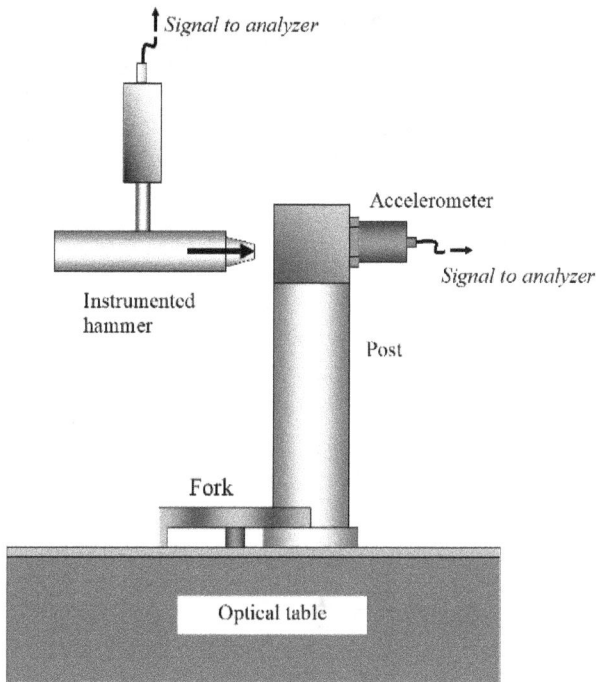

**Figure 8.3.** Test set-up for measuring dynamic compliance of an optical post [3].

posts, post holders, and mounts. This section considers an experiment specifically aimed at studying these dynamics. Let us use an optical post as an example. The test set-up consisted of an accelerometer fastened to a metal cube that was attached to the post (Fig. 8.3). As in the previous case study, the total weight of the accelerometer and the cube was close to typical weight of a mirror mount or a lens holder; all testing was conducted on the surface of a SmartTable® [1] (see Chapter 9) with active dampers to avoid the interference with table resonances.

The tests [3] compared posts of the same height (4 in., or 101.6 mm) but different diameters. Dynamic performance is considered better when the (lowest) natural frequency is higher and the corresponding resonance amplitude is lower. The test results are summarized in Table 8.1 and Fig. 8.4. The results show that the 1 in. (25.4 mm) diameter and 1.5 in. (38.1 mm) diameter posts have comparable characteristics whereas 0.5 in. (12.7 mm) diameter post is

**Table 8.1.** Test results.

| Diameter (mm) | 12.7 | 25.4 | 38.1 |
|---|---|---|---|
| Lowest resonance frequency (Hz) | 272 | 680 | 622 |
| Amplitude at resonance ($\mu$m/N) | 189 | 24.5 | 9.1 |

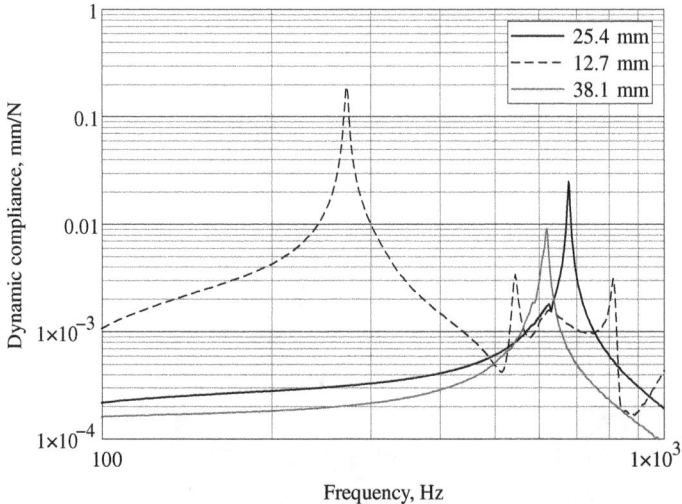

**Figure 8.4.** Dynamic compliances of optical posts with different diameters.

much more flexible with a lower frequency of resonance vibration and higher amplitude. The 1 in. post has a somewhat higher frequency, but the 1.5 in. post has a lower amplitude of resonance vibration. The lower frequency demonstrated by the 1.5 in. post can be explained by its larger mass coupled to the contact stiffness of the table surface.

These results demonstrate the importance of optomechanical element design: for example, 12.7 mm diameter post would exhibit more than seven times the displacement at a significantly lower frequency than the 25.4 mm diameter post. Dynamic compliance can be directly linked to the performance of sensitive optomechanical applications such as optical delay lines [3]. Although this series of tests was limited to one height (101.6 mm), it should be noted that posts of shorter heights provide better dynamic performance.

To summarize the lessons learned from these two case studies and similar research, the following recommendations can be made for vibration-sensitive optical equipment placed on isolated structures such as optical tables:

- Use stiff optomechanical elements; large diameter posts are preferred.
- Keep the optical axes as low over the table surface as possible.
- Use tables incorporating passive or active vibration damping.
- Place the most sensitive equipment close to the center of the table.

The next section gives a more detailed analysis of dynamic qualities of optomechanical assemblies.

## 8.4    Case Study: Dynamic Properties and Comparative Evaluation of Optical Post–Post Holder Systems

Optical post–post holder systems are ubiquitous elements of optomechanical setups. They can support optical devices such as mirrors in mirror mounts and lenses in lens holders at required positions relative to an optical bench. Analysis of specific post–post holder systems presented in this section can serve as a template for dynamic analyses of other low-damped optomechanical structures. Dynamic properties of these assemblies are vital for beam stability. Structural resonances can lead to deviation of optical planes and axes from their nominal positions and, eventually, to changes in OPL and misalignment of beams. These properties can vary substantially depending on composition, geometry, and mode of attachment. The dynamic properties of optomechanical assemblies depend also on the contact stiffness of the interface with the optical bench and the structure of the bench itself. This interface creates boundary conditions that hardly ever follow the ideal "clamped" property. That is why it is important to test optomechanical assemblies in realistic conditions, for example, on the surface of an optical table.

In this section, we compare experimentally the dynamic properties of six different post holder–post assemblies. Testing was done with three models of post holders at two different heights of the measurement axis: 152.4 mm and 101.6 mm. In all cases post holders were made of aluminum; posts were made of steel. As in previous case studies, instead of an optical assembly, a simulated payload that included an accelerometer was attached to the post; the weight of this payload was roughly equivalent to that of a typical optical mount. In some configurations an optical fork was used for attachment to the

**Table 8.2.** Post–post holder configurations under test.

| | | Composition | | | |
|---|---|---|---|---|---|
| | | Post Holder | | Post | |
| Config. no. | Diameter | Height | Connection | Diameter | Height |
| **Measurement axis height: 152.4 mm** | | | | | |
| 1 | 25.4 mm | 76.2 mm | Steel pedestal | 12.7 mm | 100 mm |
| 3 | 25.4 mm | 76.2 mm | Fork | 12.7 mm | 100 mm |
| 5 | 37.8 mm | 76.2 mm | Fork | 25.4 mm | 100 mm |
| **Measurement axis height: 101.6 mm** | | | | | |
| 2 | 25.4 mm | 76.2 mm | Steel pedestal | 12.7 mm | 50 mm |
| 4 | 25.4 mm | 76.2 mm | Fork | 12.7 mm | 50 mm |
| 6 | 37.8 mm | 76.2 mm | Fork | 25.4 mm | 75 mm |

table; in others, a pedestal was used. Table 8.2 lists the configurations subjected to dynamic testing.

The experiment followed the usual dynamic compliance test procedure. The assembly was installed at the center of an optical table with active damping (see Section 9.2 in Chapter 9) in order to eliminate the influence of the table resonances as illustrated in Fig. 8.2. The vibration sensor (accelerometer) was attached to the post via a steel cube and a stainless-steel screw. The total mass of the cube, the accelerometer, and the screw equaled 155 g. Figure 8.5 shows the test schematic. Figure 8.6 shows a photo of one of the assemblies under test. The post holder screws were aligned with the fork and the measurement axis. The peak applied force was between 7.3 N and 8.9 N in all measurements; the force profile was close to triangular, with a duration of approximately 0.01 s.

The experimental dynamic compliances for 152.4 mm high configurations are presented in Fig. 8.7. Figure 8.8 shows dynamic reactions to the force excitation by the hammer, attenuating in the time domain. Figures 8.9 and 8.10 show the same data for 101.6 mm high configurations.

A review of the experimental data presented above reveals advantages and disadvantages of each model. There are three intuitive criteria of dynamic performance:

(1) Resonance frequency, $f_n$, reflecting the stiffness-to-mass ratio — the higher the better.

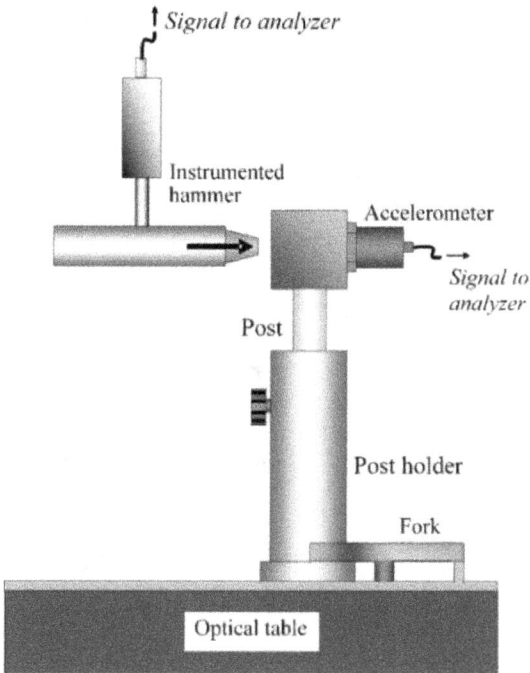

**Figure 8.5.** Test schematics for measuring dynamic compliance of an optical post–post holder system.

(2) Resonance amplitude, $A_n$, reflecting the maximum vibration amplitude per unit force — the lower the better.

(3) Attenuation rate, defining the settling time — the higher attenuation (shorter time) the better.

Different criteria can lead to different preferences. For example, structures with higher resonance frequencies may feature higher amplitudes. This issue has been addressed in the previous chapter in relation to the dynamic performance of optical tabletops. In this case, Dynamic Deflection Coefficient (DDC) and Maximum Relative Motion (MRM) criteria were introduced that combine natural frequencies and resonance amplitudes (see Section 7.3 of Chapter 7).

To introduce the combined criteria for this case, refer to the basic dynamic equations describing the vibration of the system as a single degree of freedom oscillator (a mass-spring system on a rigid base)

**Figure 8.6.** Assembly under test. Configuration no. 6.

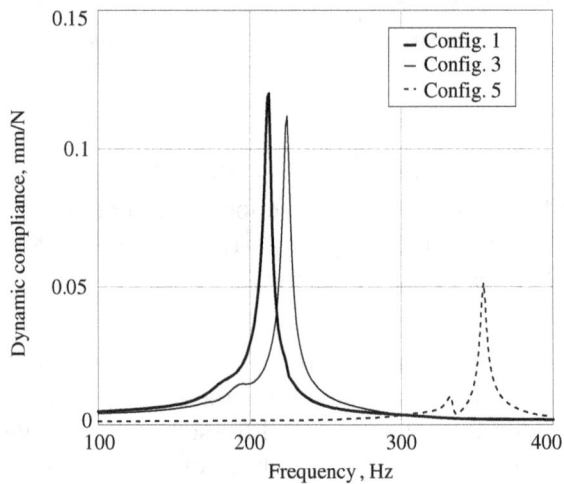

**Figure 8.7.** Comparison of dynamic compliances. Measurement axis at 152.4 mm height.

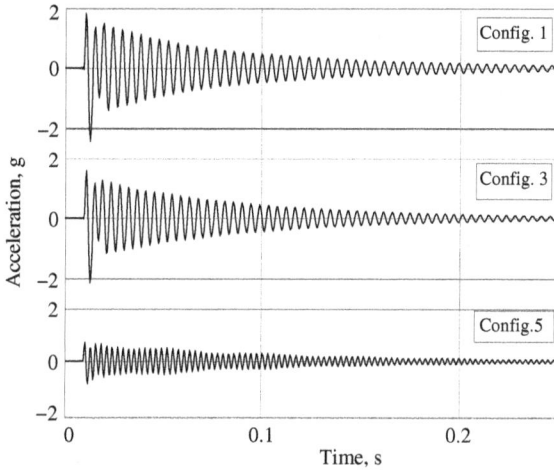

**Figure 8.8.** Vibration attenuation after an impact. Measurement axis at 152.4 mm height.

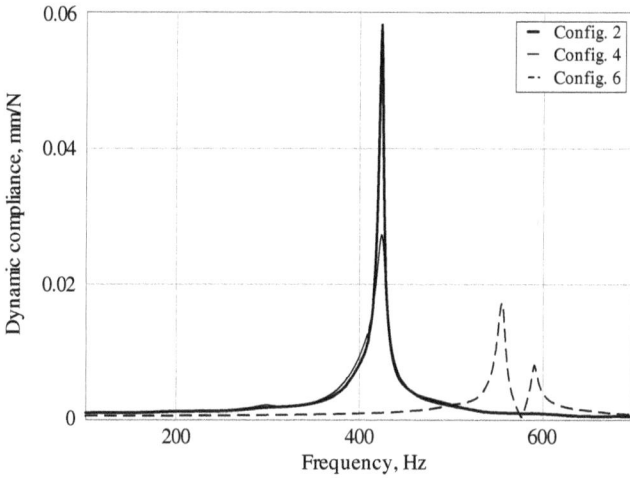

**Figure 8.9.** Comparison of dynamic compliances. Measurement axis at 101.6 mm height.

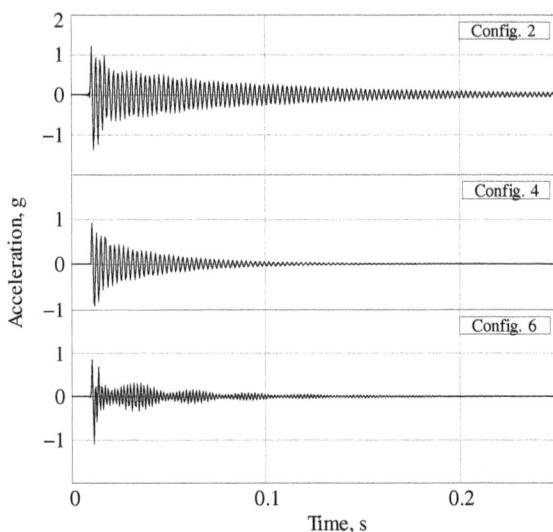

**Figure 8.10.** Vibration attenuation after an impact. Measurement axis at 101.6 mm height.

described in Chapter 3:

$$\ddot{x} + \frac{c}{m}\dot{x} + \frac{k}{m}x = \frac{F(t)}{m}, \tag{8.1}$$

where $x$ is the displacement of the mass, $m$, $c$ and $k$ are the mass, damping, and stiffness coefficient, respectively, $F(t)$ is a force excitation on the mass. This equation is similar to the equation for relative motion of the mass when the base is subjected to prescribed vibration, $y$:

$$\ddot{z} + \frac{c}{m}\dot{z} + \frac{k}{m}z = -\ddot{y}, \tag{8.2}$$

where $z = x - y$.

Defining the natural frequency by

$$f_n = \frac{1}{2\pi}\sqrt{\frac{k}{m}}$$

and the loss factor by

$$\eta = c/\sqrt{km},$$

we can write both equations in the generic form:

$$\ddot{r} + 2\pi\eta f_n \dot{r} + (2\pi f_n)^2 r = s(t).$$

The response of the system in the frequency domain is described by a transfer function, or frequency response function, $H(f)$, with a magnitude given by the ratio of power spectral densities of the output and the input:

$$|H(f)|^2 = \frac{W_r(f)}{W_s(f)} = \frac{1}{(2\pi f_n)^2 \left[\left(1 - \frac{f^2}{f_n^2}\right)^2 + \left(\eta\frac{f}{f_n}\right)^2\right]}.$$

For a broad-band stationary random source, if $\eta \ll 1$, so that the response is dominated by the resonance zone, the mean-square response value is determined by

$$\langle r \rangle^2 = W_s(f_n) \int_0^\infty |H(f)|^2 df = \frac{W_s(f_n)}{32\pi^3 \eta f_n^3}. \tag{8.3}$$

Considering the continuous systems such as post holder–post systems under test, the above equations are still applicable to the modal response of the system (see Section 3.5 of Chapter 3), that is, the response due to one vibration mode, or one resonance frequency. In this case, $m$ represents the "equivalent" mass of the mode, which is, along with the natural frequency and modal loss factor, determined from experimental data. The dynamic compliance is expressed in the vicinity of the resonance by the ratio of the complex amplitude of the displacement at the measurement point, to the complex amplitude of the force:

$$G(f) = \frac{x}{F} = \frac{1}{m} \frac{1}{(2\pi f_n)^2 \left(1 - \frac{f^2}{f_n^2} + i\eta\frac{f}{f_n}\right)} = \frac{1}{m} H(f).$$

The absolute value of the compliance is given by

$$|G(f)| = \frac{1}{m} \frac{1}{(2\pi f_n)^2 \sqrt{\left(1 - \frac{f^2}{f_n^2}\right)^2 + \left(\eta\frac{f}{f_n}\right)^2}}.$$

The natural frequency can be estimated from the experimental dynamic response curve directly by the maximum point of compliance:

$$G_{\max} = G(f_n). \tag{8.4}$$

The loss factor can be estimated from experimental data by the "half-power width" of the resonance curve [4]:

$$\eta = \Delta f / f_n, \tag{8.5}$$

where $\Delta f$ is the difference between two frequencies below and above resonance with $G(f) = G_{\max}/\sqrt{2}$. Smooth interpolation of experimental data was performed for precise application of Eqs. (8.4) and (8.5). The equivalent mass is then calculated as

$$m = \frac{1}{G_{\max}\eta(2\pi f_n)^2}. \tag{8.6}$$

As a result, the RMS deflection of the post in response to random force excitation can be estimated from experimental data according to (8.3) and (8.6):

$$x_{\text{RMS}} = \frac{1}{m}\frac{\sqrt{W_F(f_n)}}{\sqrt{32\pi^3 f_n^3 \eta}} = \sqrt{\frac{\pi}{2}}G_{\max}\sqrt{\eta f_n}\sqrt{W_F(f_n)}.$$

Therefore, RMS absolute deflection of the post per unit of square root power spectral density of force,

$$x_{\text{RMS}}^0 = \frac{1}{m\sqrt{32\pi^3 f_n^3 \eta}} = \sqrt{\frac{\pi}{2}}G_{\max}\sqrt{\eta f_n}, \tag{8.7}$$

can serve as a figure of merit for the post holder–post structure.

Two other figures of merit can be considered. If we are mostly interested in the reaction of the structure to the motion of the attachment point, we can estimate its relative motion in response to the base acceleration by the following expression based on the model (8.2):

$$z_{\text{RMS}} = \frac{\sqrt{W_{\ddot{y}}(f_n)}}{\sqrt{32\pi^3 f_n^3 \eta}}.$$

This leads to the dynamic deflection coefficient (DDC) similar to that used in ranking optical tables and breadboards (see Section 7.3 of Chapter 7):

$$\text{DDC} = \frac{1}{\sqrt{f_n^3 \eta}}. \tag{8.8}$$

Finally, in some applications the speed with which the structure recovers from a disturbance (settling time) is important. As discussed above in Section 3.2 of Chapter 3, the solution of Eq. (8.1) after an initial pulse is a sinusoidal oscillation with amplitudes attenuating as $\exp(-\pi\eta f_n t)$. Hence the attenuation rate,

$$\kappa = \pi\eta f_n, \qquad (8.9)$$

can be another figure of merit. Note that, contrary to criteria (8.7) and (8.8), higher values of criterion $\kappa$ are preferable. Out of the three criteria introduced above, criterion (8.7) seems to be most comprehensive since it reflects all three physical characteristics of the structure: mass, stiffness-to-mass ratio, and damping.

Natural frequencies, peak amplitudes, loss factors, and all three figures of merit for six configurations of post holders under test are summarized in Table 8.3. Most of the experimental data obtained during this test indicate essentially unimodal response, that is, the motion dominated by one mode. Two modes were considered in configurations 5 and 6. In these cases, mean square values add up; therefore, the DDC and RMS calculated for each mode were summed by the Pythagorean rule. The attenuation rate shown is the lower of the two modes.

**Table 8.3.**  Dynamic characteristics of different post holder–post assemblies.

| | Measurement Axis Height: 152.4 mm | | | Measurement Axis Height: 101.6 mm | | |
|---|---|---|---|---|---|---|
| Config. No. | 1 | 3 | 5 | 2 | 4 | 6 |
| Resonance frequency (Hz) | 213 | 225 | 355 | 424 | 424 | 556 |
| 2nd resonance frequency (Hz) | | | 333 | | | 591 |
| Loss factor | 0.022 | 0.021 | 0.011 | 0.010 | 0.026 | 0.015 |
| 2nd loss factor | | | 0.016 | | | 0.013 |
| Spectral amplitude (mm/N) | 0.121 | 0.112 | 0.051 | 0.057 | 0.027 | 0.017 |
| 2nd spectral amplitude (mm/N) | | | 0.010 | | | 0.008 |
| Dynamic Deflection Coefficient, $s^{3/2}$ (8.8) | 0.0022 | 0.0020 | 0.0021 | 0.0011 | 0.0007 | 0.0012 |
| Attenuation rate, 1/s, minimum (8.9) | 14.4 | 15.0 | 12.3 | 13.9 | 34.2 | 23.7 |
| RMS deflection, mm/(N/$\sqrt{\text{Hz}}$) (8.7) | 0.32 | 0.31 | 0.13 | 0.15 | 0.11 | 0.09 |
| Rank | 6 | 5 | 3 | 4 | 2 | 1 |

In the last row, the post holders are ranked according to the criterion (8.7). As expected, configurations with the lower measurement axis (101.6 mm) vastly outperform corresponding designs with the higher axis (152.4 mm). For example, configurations 1 and 3 would produce about twice higher resonance displacement compared to configurations 2 and 4 using the same post holders at lower axis for the same disturbance.

Again, this case study confirms the general recommendation to keep the optical axis as close as possible to the bench. For the same height, larger post holders (configurations 5 and 6) are preferable; configurations using forks slightly outperform configurations using pedestals.

## 8.5   Case Study: Integral Damping in a Mirror Mount Interface

Our next study is an experiment on vibration damping of a mirror mount. High-frequency vibration of a mirror mount (shown in Fig. 8.11) was noticed in an external cavity diode laser application.

The resonance response of the mirror can be reduced by introducing dissipation of mechanical energy (damping) into the support structure as explained in Chapters 3 and 7. A series of experiments was conducted with damping attempted by using thin elastomeric washers at the interface between the mount and the supporting structure (the post). In each case, vibration transmissibility was measured from the moving platform to the middle of the circular aluminum plate installed into the mirror mount and imitating the mirror (see Fig. 8.12).

The results are presented in Fig 8.13. The graphs plot transmissibility vs. frequency on a linear scale. In absence of washers, the mount exhibited a very well-defined high amplitude main resonance at 896 Hz. This resonance is created mainly by the limited stiffness of the contact area. Application of elastomeric washers reduced the resonance response drastically. Three curves, corresponding to three washers of different thicknesses and different elastomeric materials, represent different degrees of the trade-off between damping and stiffness. Unfortunately, there is always a trade-off: thin layers of elastomer, although very stiff in the transversal direction, are resilient in

**Figure 8.11.** Mirror mount U100-AC.

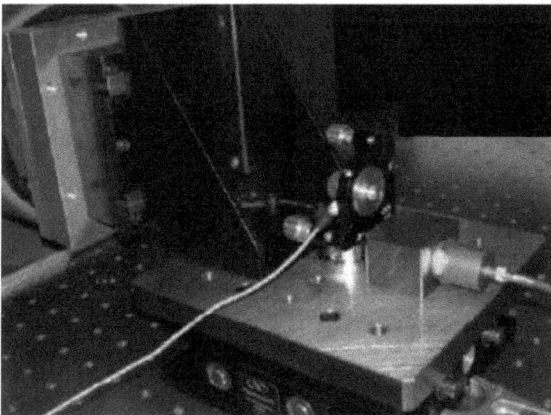

**Figure 8.12.** Test set-up for measuring resonance properties of a mirror mount assembly. The mirror mount is installed, through a post, on a movable platform that is excited by an electrodynamic shaker. An accelerometer is fastened to an aluminum circular plate installed in place of a mirror. A reference accelerometer is installed on the platform.

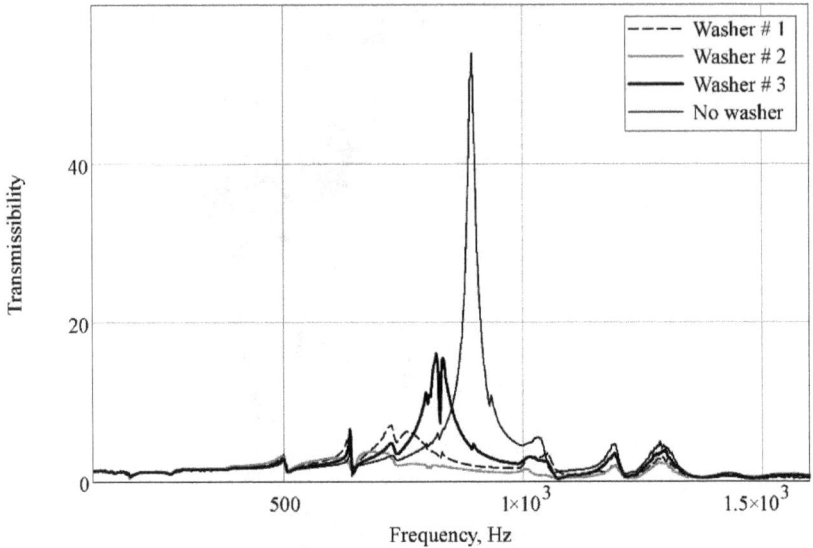

**Figure 8.13.**   Resonance attenuation by elastomeric washers.

the tangential direction, which can lead to a reduction in the overall structural stiffness of the assembly. Thermal stability may be a concern, too.

The next two case studies describe troubleshooting of vibration problems. Troubleshooting becomes necessary when the problem is discovered unexpectedly at a late stage of product design, optical test, or system integration. This may be a difficult and time-consuming process that should be avoided if possible. The earlier in the design process vibration control considerations are addressed, the better. Troubleshooting of vibration problems usually involves vibration measurements, model-based hypotheses, and test-based conclusions.

## 8.6   Case Study: Fans and Lasers

In this case, increased vibration of an ultrafast laser, affecting its performance, was noticed. The laser was installed on an optical table. The cooling fan in the laser enclosure had the nominal rotation speed of 5200 RPM, or 86.7 Hz. If it were the fan that caused vibration, this frequency was expected to be present in the vibration

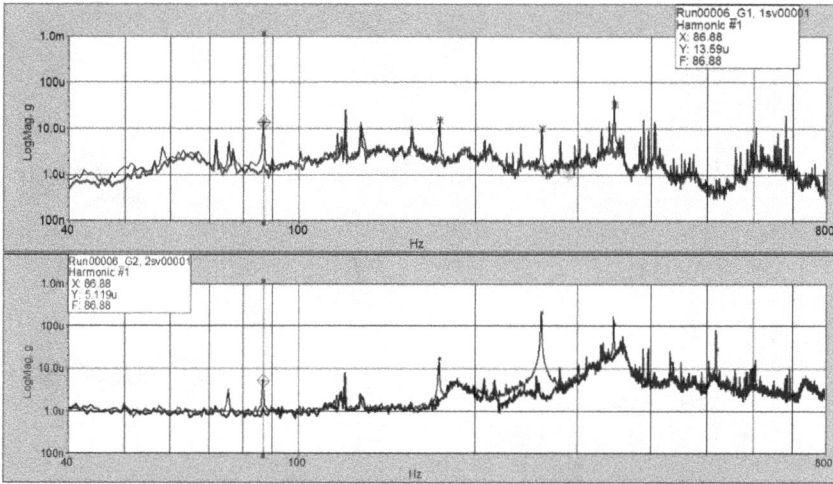

**Figure 8.14.** Vibration spectra. Top graph — on the surface of the table; bottom graph — on the laser casing in the horizontal direction. Thin lines — fans powered, thick lines — power off. Peaks at the rotation frequency and harmonics are dominating the spectrum.

spectra along with its harmonics at 173.4 Hz, 260.1 Hz, 346.8 Hz, 433.5 Hz, etc.

Figure 8.14 compares vibration spectra measured with accelerometers installed on the surface of the table (top graph) and on the wall of the laser casing (bottom graph). The frequency (86.9 Hz) very close to the nominal rotation frequency of the fan, and its harmonics are clearly present at both locations. This result shows that the contribution of the fans to the laser casing and table surface vibration is indeed very significant and even dominating.

The decision was made to arrange vibration isolation of the fans by increasing the mass of their supporting plates and introducing pieces of soft acoustical foam as isolators. The next set of test data (Fig. 8.15) shows that the isolation reduces frequency peaks associated with fans to the background level. It should be noted that the "fans unplugged" and "fans isolated" data were taken several hours apart, and the acoustical environment changed during this time. The peaks around 60 Hz and 120 Hz are results of the changing environment, not side effects of isolation. (This was confirmed by a separate test.)

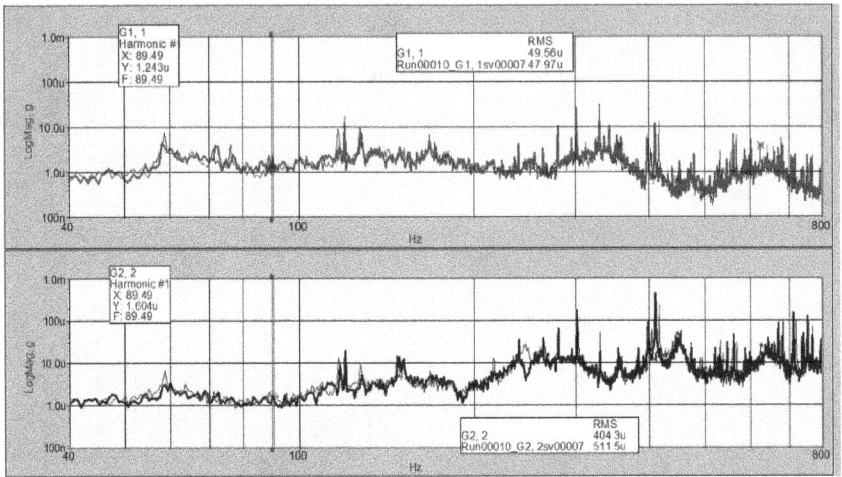

**Figure 8.15.** Vibration spectra with fans isolated (thin lines) vs. fans disconnected (thick lines, taken a few hours earlier). Top graph — on the surface of the table, bottom graph — on the laser casing in the vertical direction. Vibration levels with fans isolated are nearly identical to that with fans completely disconnected. Peaks around 60 and 120 Hz are due to the environment.

The fans as sources of vibration were found out and excluded by this troubleshooting process.

## 8.7 Case Study: A Vibrating Camera

This case study describes vibration troubleshooting of an assembly and test workstation. The schematic of the workstation is shown in Fig. 8.16.

The test workstation was designed to support the build of prototype lens assemblies. Lens stacks included several lenses of varying diameters. The stacks could stand rather high, which dictated the height of the frame. The test consisted of interferometer measurements of WFE (wavefront error) using double-pass geometry. The goal was to achieve RMS WFE of less than 0.025 waves. The interferometer was not able to capture measurements due to vibrations. Vibration troubleshooting was necessary to find the sources of this vulnerability and plan the vibration mitigation steps.

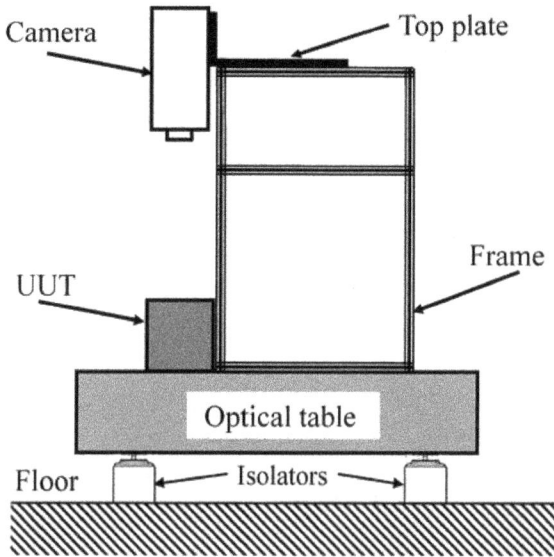

**Figure 8.16.** Schematic of the test workstation. UUT (unit under test) may include a large multi-lens stack.

The strategy of the troubleshooting was to make two-channel measurements with two accelerometers, trying to identify resonance properties and weak links in the structure. The graphs in Figs. 8.17–8.22 show vibration transmissibilities on the top charts and linear spectra ($\Delta f = 0.25$ Hz) of both channels on the bottom charts.

First, to check the performance of the vibration isolators supporting the test rig, we place one accelerometer on the floor, another on the platform (Fig. 8.17(a)). The isolators were so-called air springs, a basic form of pneumatic vibration isolators combining a rubber shell and air volume. The performance of air spring isolators is as expected, with the natural frequency of about 4.25 Hz. Two-chamber pneumatic isolators (see Section 6.4 of Chapter 6) would provide lower natural frequencies, better overall isolation, and higher damping.

Second, we probe into the properties of the upper structure by putting channel 1 accelerometer on the platform and channel 2 on the top plate of the frame, both in the vertical direction (Fig. 8.18). Not much difference is noticed between the vibration levels at these two points up to the main resonance frequency of the frame at about 70 Hz. So the frame proved to be surprisingly stiff.

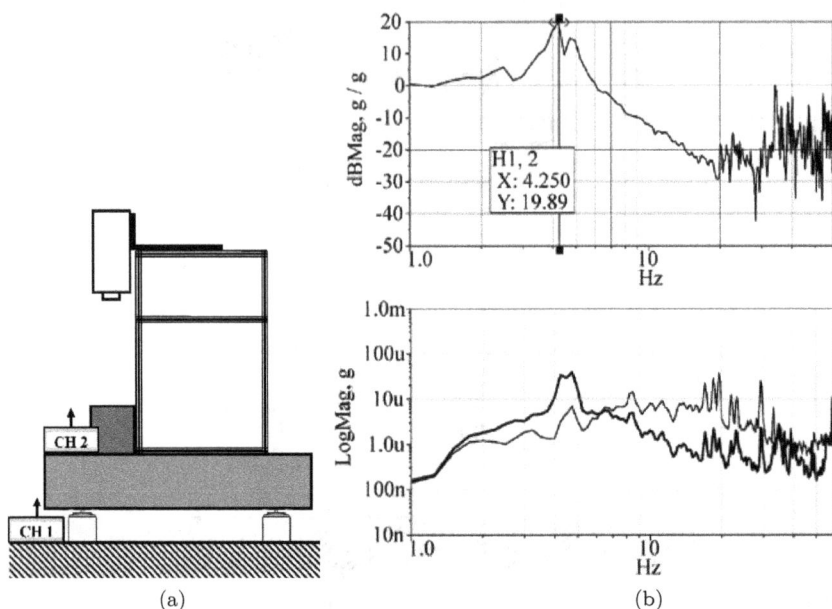

**Figure 8.17.** Test setup for estimating performance of vibration isolators (a) and results (b): top — transmissibility, bottom — spectra (thin line — channel 1, thick line — channel 2).

Next, the channel 1 accelerometer stays on the platform, and the channel 2 accelerometer is placed on top of the camera (Fig. 8.19(a)). Now another low-frequency peak becomes prominent at about 20 Hz (Fig. 8.19(b)).

This resonance is seen even more clearly if the reference accelerometer (channel 1) is placed on the interface plate (Fig. 8.20). The same resonance frequency is indicated by measuring horizontal accelerations in the plane of symmetry of the structure as shown in Fig. 8.21.

Measurements in transversal horizontal direction reveal another vibration mode with a slightly higher resonance frequency (Fig. 8.22).

This series of measurements led to clear conclusions and recommendations (see Fig. 8.23). First, there is a possibility to use better isolators. Second, most importantly, the two-channel measurements found a weak link in the structure: insufficient stiffness of attachment of the camera to the frame gave rise to low-frequency high amplitude

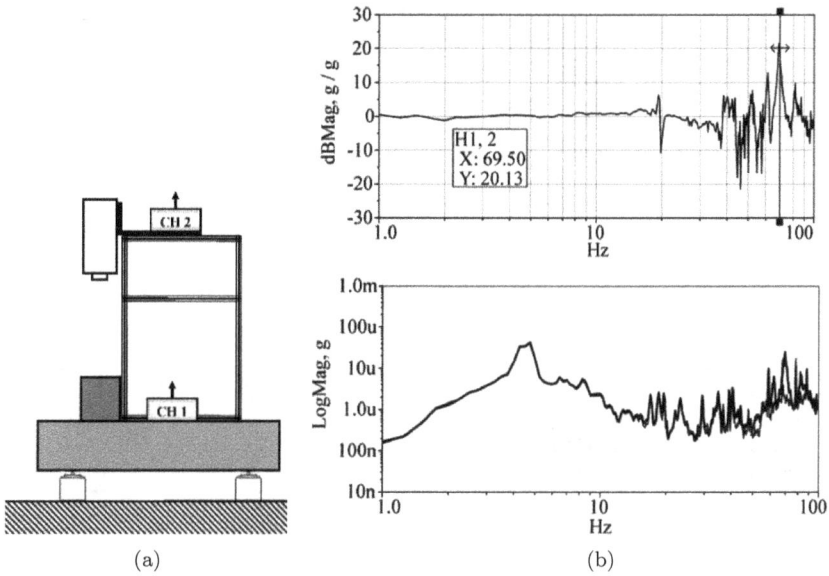

**Figure 8.18.** Test setup for estimating resonance properties of the frame (a) and results (b): top — transmissibility, bottom — spectra (thin line — channel 1, thick line — channel 2).

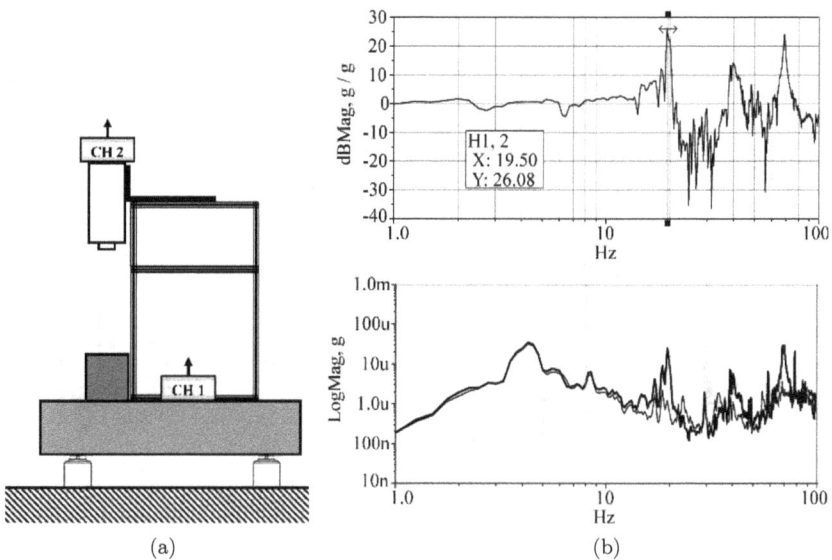

**Figure 8.19.** Test setup for estimating resonance properties of the frame and the bracket (a) and results (b): top — transmissibility, bottom — spectra (thin line — channel 1, thick line — channel 2).

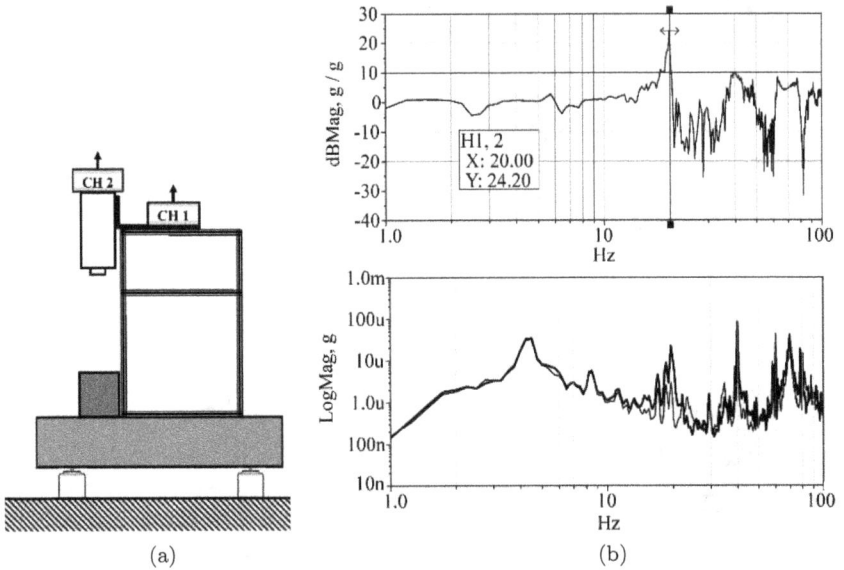

**Figure 8.20.** Test setup for estimating resonance properties of the bracket (a) and results (b): top — transmissibility, bottom — spectra (thin line — channel 1, thick line — channel 2).

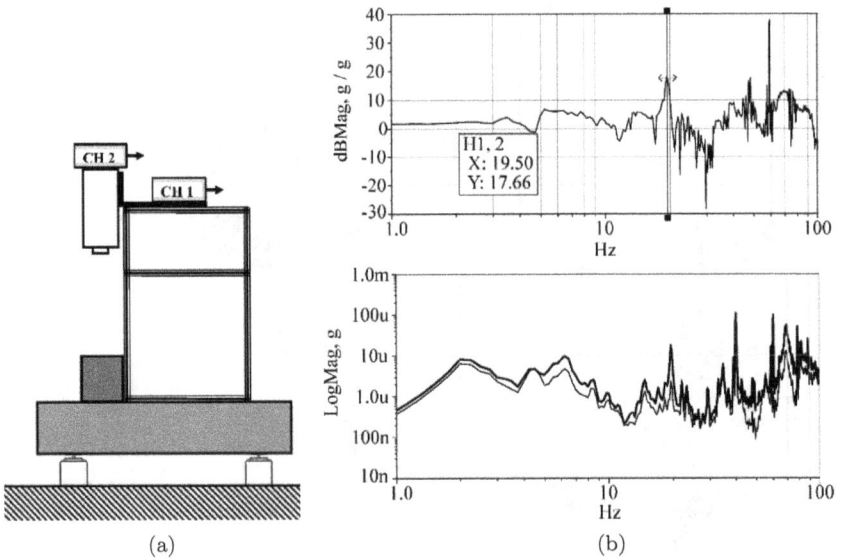

**Figure 8.21.** Test setup for estimating resonance properties of the bracket by accelerometer measurement in the horizontal direction (a) and results (b): top — transmissibility, bottom — spectra (thin line — channel 1, thick line — channel 2).

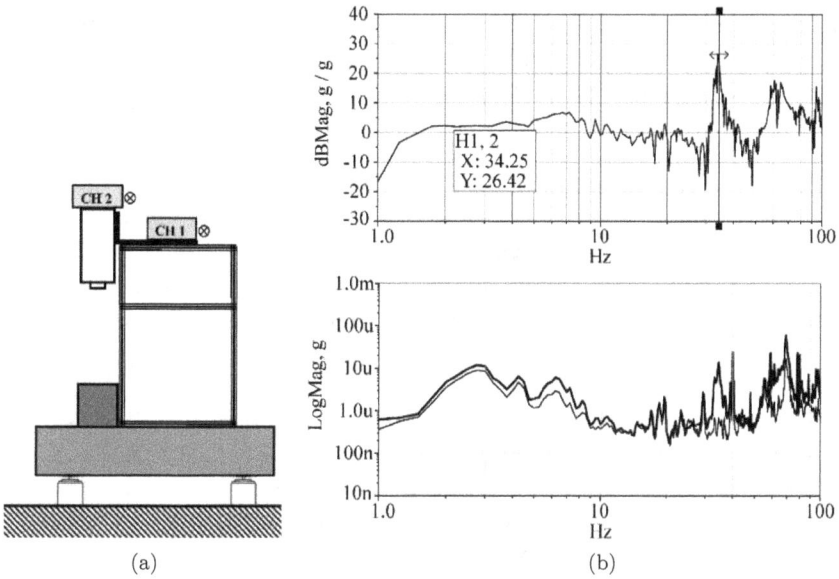

**Figure 8.22.** Test setup for estimating resonance properties of the bracket (a) by accelerometer measurement in horizontal direction $y$ and results (b): top — transmissibility, bottom — spectra (thin line — channel 1, thick line — channel 2).

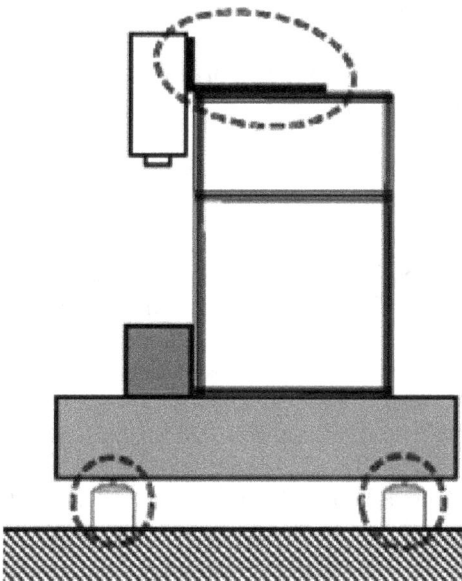

**Figure 8.23.** Areas of improvement identified by troubleshooting.

resonance vibration. A stiffer connection was needed here. Implementation of these two recommendations made high-quality interferometry measurements possible.


## 8.8   Lessons Learned

In summary, the lessons learned through the case studies lead to the following general guidelines for troubleshooting vibration issues and improving the resistance of optomechanical assemblies and applications to vibration.

- Identify the sources of vibration through a vibration survey.
- Try to isolate the sources if possible.
- Identify the weak points in dynamic properties of the structure through two-channel measurement or analysis.
- Correct, if possible, by increasing the mechanical stiffness and introducing dissipative elements.
- Do not forget even the smallest part of the assembly.
- You may need to apply *ad hoc* solutions such as dynamic absorbers or damping pads, or resort to active methods of vibration control (Chapter 9).

The earlier in the design process vibration control considerations are taken into account, the better.


## References

1. V.M. Ryaboy, P.S. Kasturi, A.S. Nastase, and T.K. Rigney, Optical table with embedded active vibration dampers (Smart table), *Proceedings of SPIE Vol. 5762, Smart Structures and Materials 2005: Industrial and Commercial Applications of Smart Structures Technologies*, edited by E.V. White, pp. 236–244.
2. J. Fisher and V. Ryaboy, Effective vibration reduction stabilizes laser beams, *Laser Focus World*, August 2008, pp. 92–97.
3. V.M. Ryaboy and J. Fisher, Vibration isolation in optical test systems, *Photonics Tech Briefs*, September 2011, pp. 92–95.
4. R.G. DeJong, Statistical methods for analyzing vibrating systems, C.M. Harris (ed.), in *Shock and Vibration Handbook*, 4th edn., McGraw-Hill, New York, 1995, p.11.25, and subsequent editions.

# Chapter 9

# Active Vibration Control

This chapter is devoted to systems with enhanced performance and increased complexity compared to the devices studied so far, namely, the active systems that are becoming more and more widely used for vibration control of precisions optomechanical systems. In the last section we'll review one-of-a-kind extremely effective active vibration control arrangements that enabled unique experiments and major scientific breakthroughs.

Active vibration control systems use real-time information from vibration sensors. They include control systems and external sources of energy to create forces with proper amplitude and phase to reduce unwanted vibration.

As a terminological note, the reader must have in mind that in older literature the term "active" was sometimes applied to vibration control systems that adjust their parameters in response to changing conditions. A familiar example would be a pneumatic isolator with a leveling valve, where the stiffness and load capacity of the isolator are adjusted to the supporting load via positional feedback; another example is a shock system for a vehicle that can adjust damping and stiffness to the terrain. The proper term for this kind of system is "semi-active" system. We'll reserve the term "active system" for the systems that react in real time to an offending vibration signal with a mitigating vibration signal.

The three necessary components of an active vibration control system are vibration sensors, control circuits, and actuators.

The literature on active vibration control is enormous. Fundamental monographs [1–3] can serve as a general introduction to this subject. This chapter reviews only methods and design schematics that found applications to precision vibration control of optomechanical systems. Presently, active vibration control systems gain increasing recognition in optomechanical applications. Higher complexity and cost are the main obstacles to wider acceptance. However, they seem unavoidable in advanced applications that push the boundaries of achievable precision and sensitivity.

It had been proven [4, 5] that passive isolation systems have an inherent limitation on their mass and static stability that are valid for the whole class of systems, no matter how complex they are. In practice, those limitations restrict the achievable performance of passive systems at low (single Hz range) frequencies. Active systems can overcome these restrictions. They have certain general limitations in terms of work performed by actuators [5–7], but these are not critical for precision applications.

## 9.1   Active Isolation

Active isolators, quite like passive isolators, are placed between the source of vibration and the protected object, and their goal is to prevent vibration emitted by the source from reaching the object. Active vibration isolators generally fall into two categories: stiff (sequential) and soft (parallel) systems.

Figure 9.1 shows a single-degree-of-freedom (SDOF) schematics of a stiff active isolation system that found some practical applications [8,9]. Here the isolated object (payload) is modeled by a mass $m$, and the elastic element by a (possibly complex) stiffness $k$. A vibration sensor produces an electric signal that is conditioned and transformed by the analog or digital controller into a driving voltage applied to a piezoelectric actuator. Piezoelectric actuator is a stiff actuator that reacts to this voltage with elongation or contraction. The control goal is to compensate the dynamic displacement of the mass $m$ caused by the motion of the floor $u_0$ by the elongation/contraction of the piezo, thereby minimizing the resulting dynamic displacement $u$.

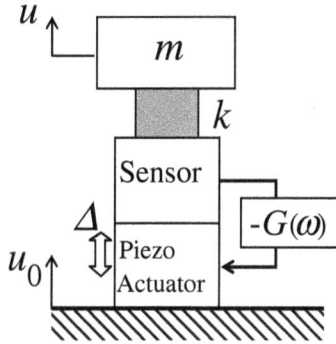

**Figure 9.1.** SDOF model of a stiff piezoelectric active vibration isolation system with sequential structure.

The equation of motion of the mass $m$ in complex amplitudes (see Section 2.2 of Chapter 2) is

$$(k - m\omega^2)u = k(u_0 + \Delta), \qquad (9.1)$$

where $\Delta$ is the elongation of the piezo. Feedback control function, $-G(\omega)$, incorporating transfer functions of the sensor and actuator, signal conditioning filters, and control algorithm, transforms displacement at the measurement point (at the top of the piezo in this case) into elongation of the piezo:

$$\Delta = -G(\omega)(u_0 + \Delta). \qquad (9.2)$$

Equations (9.1) and (9.2) lead to the following expression for the vibration transmissibility of the active isolation system:

$$\frac{u}{u_0} = \frac{k}{k - m\omega^2} \frac{1}{1 + G(\omega)}. \qquad (9.3)$$

Here the first factor on the right-hand side is the passive transmissibility; the second term describes the reduction of vibration amplitude by the active system. Equation (9.3) allows us to make important observations concerning stiff sequential active vibration isolation in general. First, the resonance response of the passive system enters as a factor in the active transmissibility, so the passive resonance will be present in the spectrum of the vibration of the isolated object, which is undesirable. The control algorithms usually attempt to correct for the passive resonance. The optimum phase of the control signal, from

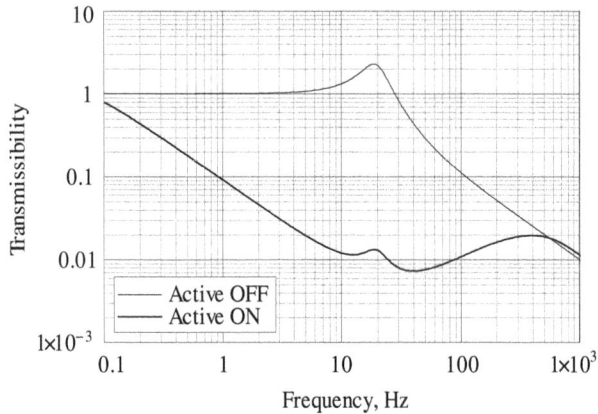

**Figure 9.2.** Theoretical transmissibility of the SDOF model of a stiff piezoelectric vibration isolator.

stability and performance standpoints, is 180°; so, the system is trying to counteract the motion of the isolated object directly. Note that in this structure the active system practically does not affect the dynamic compliance, that is, the resistance to the force acting directly on the payload. Dynamic compliance is defined by the stiffness of the elastic element that should be chosen sufficiently high. Placing the sensor directly on the isolated object would make the system performance sensitive to higher resonances that are always present in realistic systems.

Figure 9.2 shows theoretical transmissibility derived for this model based on 20 Hz passive isolator, using a geophone (see Section 4.1 of Chapter 4) as vibration sensor, and including high-pass and low-pass filters that ensure rejection of quasi-static noise and roll-off at high frequencies; control gain was restricted by the stability requirement. Passive isolation transmissibility is shown for comparison. The model predicts remarkable performance at low frequencies: transmissibility as low as 0.1 or −20 dB at 1 Hz, which is not achievable by passive systems. At higher frequencies, the performance rolls off to that of the passive isolator.

In practice, the performance is restricted by factors not included in the simple model of Fig. 9.1, such as three-dimensional character of the system and higher resonances of the payload. Figure 9.3 shows the measured transmissibility of a commercial piezoelectric active

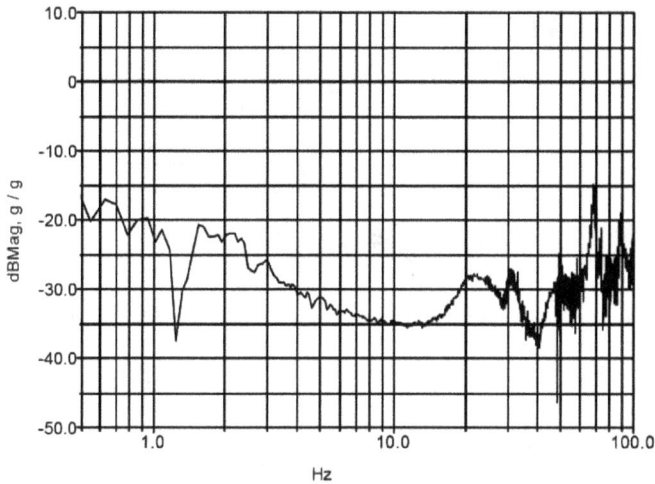

**Figure 9.3.** Measured vertical transmissibility between floor and payload motion for a stiff piezoelectric vibration isolator. Reproduced with permission from [9].

isolation system described by Anderson and Houghton [9]. Tenfold vibration reduction at 1 Hz is achieved, as in the theoretical estimate. A local maximum at the resonance frequency of the passive isolator (about 22 Hz) is noticeable, as in the theoretical prediction. The performance at higher frequencies is affected by payload resonances.

Another configuration of an active vibration isolation system is a soft electromagnetic control system working in parallel with the main vibration isolator. A simple SDOF model of such a soft (parallel) active isolation system is shown in Fig. 9.4. Here the isolated object is represented by the mass $m$; the actuator is placed in parallel with the main elastic support modeled by a (possibly complex) stiffness $k$. The equation of motion in complex amplitudes is

$$(k - m\omega^2)u = ku_0 + F, \qquad (9.4)$$

where $F$ is the force generated by the actuator. Feedback control function, $-G(\omega)$, incorporating transfer functions of the sensor and actuator, signal conditioning filters, and control algorithm, transforms displacement at the measurement point (connected directly to the isolated object in this case) into the force generated by the

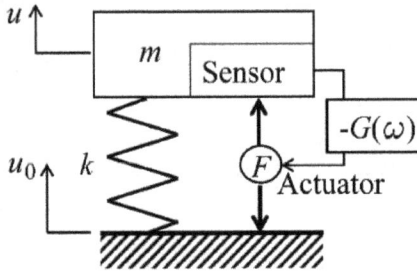

**Figure 9.4.** SDOF model of a soft electromagnetic active vibration isolation system with parallel structure.

actuator:

$$F = -G(\omega)u. \tag{9.5}$$

Equations (9.4) and (9.5) lead to the following expression for the vibration transmissibility of the active isolation system:

$$\frac{u}{u_0} = \frac{k}{k - m\omega^2} \frac{1}{1 + \frac{G(\omega)}{k - m\omega^2}}. \tag{9.6}$$

Here, again, the first factor on the right-hand side is the passive transmissibility; the second term describes the reduction of vibration amplitude by the active system. Equation (9.6) shows that in this case the resonance response of the main system becomes, essentially, a part of the control effort: the higher the resonance reaction, the higher is the control action. This is true also for the higher resonances of the payload, which makes the system less sensitive to the resonance properties of the payload despite the sensor being placed directly on the isolated object. The phase of the feedback function, $-G(\omega)$, is adjusted to the frequency response of the passive system; at the resonance, the optimum phase is $-90°$. The active system affects not only the transmissibility but also the dynamic stiffness. The feedback gain is limited by conditions of stability that stem from realistic frequency responses of the sensor and actuator, as well as necessary low-pass and high pass filters in the control electronics. Figure 9.5 shows an example of projected performance based on this simplified model and illustrates tradeoffs between isolation effect, off-range amplification, and stability margin, driven by the feedback gain. As feedback gain increases, isolation performance increases, but

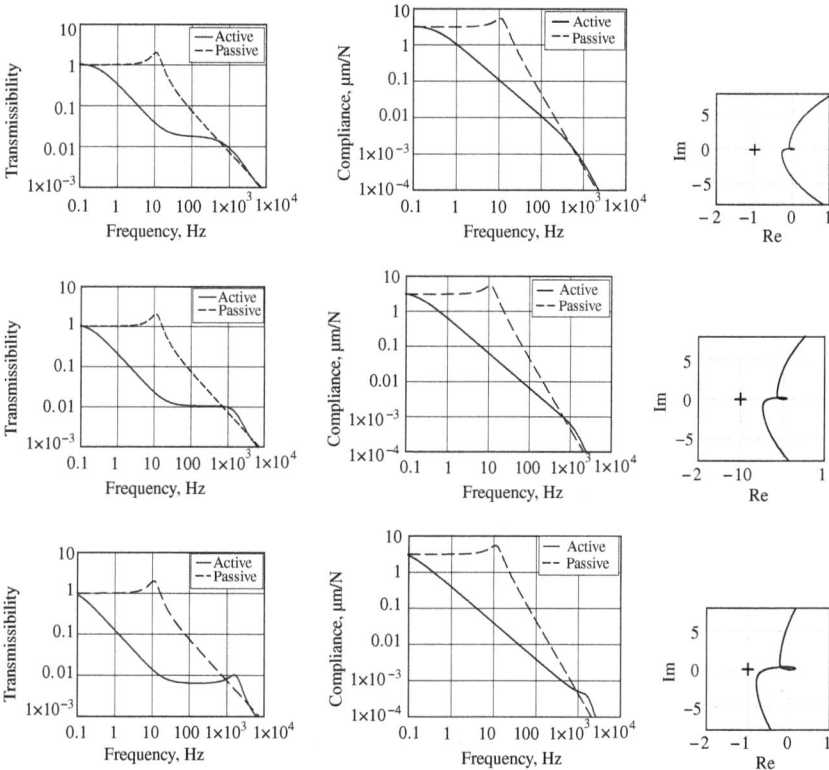

**Figure 9.5.** Projected performance and setup tradeoffs for the SDOF model of a parallel-structure active electromagnetic vibration isolator, illustrated by vibration transmissibility, dynamic compliance, and Nyquist diagram. Top — low gain, middle — medium gain, bottom — high gain. Solid lines — active, dashed lines — passive. Load 60 kg.

side effects of amplification outside the frequency range occur, and stability margin decreases as the Nyquist diagram comes closer to encompassing the point $(-1, 0)$.

The real-life performance of electromagnetic active vibration isolation systems in 6-DOF is limited by cross-axis interactions, frequency responses of actuators and sensors, and electronic noise. An example of practically achieved performance of a commercial 6-DOF electromagnetic active vibration isolation system [10] based on electromagnetic actuators working in parallel with a spring-based passive

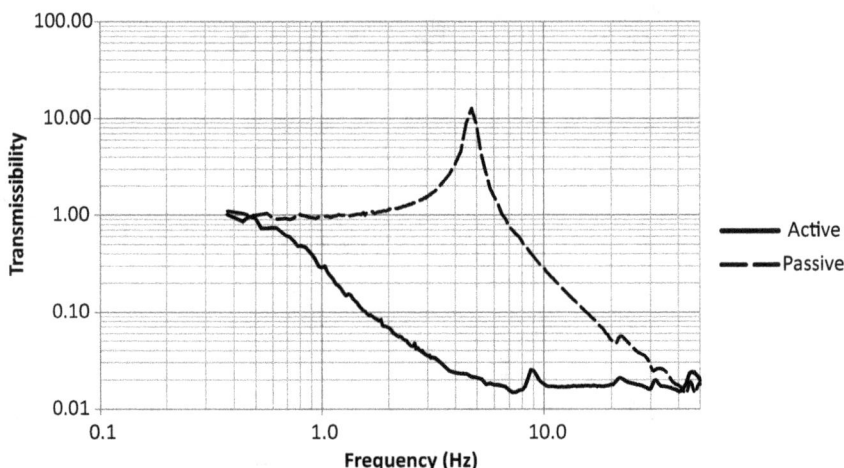

**Figure 9.6.**   Transmissibility of the 6-DOF electromagnetic active vibration isolation workstation in vertical direction [10]. Permission to use granted by Newport Corporation. All rights reserved.

isolation system is shown in Fig. 9.6. An example of active electromagnetic vibration control acting in parallel with a passive pneumatic system can be found in [11].

More intricated control schematics can be used for further improving the isolation. In particular, using additional feedforward controls from motion sensors on the foundation can improve stability; signals from relative motion sensors can be used to reduce the effects of the sensor noise. Examples of advanced active vibration isolation systems are considered below in Section 9.3.

## 9.2   Active Damping

Even if excellent protection from floor vibration is achieved by vibration isolators, the isolated platform itself deviates from the ideal rigid-body behavior at natural frequencies of its flexural vibrations. These flexural vibrations cause misalignment of optical equipment installed on the isolated platform, which can lead to deterioration of optical performance.

Various known passive means of reducing these unwanted vibrations, such as structural damping described in Chapter 7, have only limited effect or require tuning to particular resonance frequencies. Methods of active vibration control offer a promise of high efficiency without the restrictions of passive systems. Active vibration dampers have the same basic elements as active vibration isolators, but they serve a different goal. The physical mechanism they implement is energy dissipation as opposed to direct suppression of vibrations by counteracting forces. Active dampers do not support the load. Active damping is effectively used for suppression of higher-frequencies structural resonances of extended structures.

Briefly consider the theoretical background of active vibration damping following the paper [12]. For a general active multi-degree-of-freedom system, dynamic deflection at an arbitrary point, $u_n$, caused by an external excitation force, $f_m$, can be represented in the frequency domain by

$$u_n = G_{nm}(\omega) \cdot f_m + \sum_{l=1}^{L} G_{nl}(\omega) \cdot f_l^{act}, \qquad (9.7)$$

where $f_l^{act}$ are active forces created by control loops, and $G_{nm}(\omega)$ are dynamic compliances. It has been established [1, 13] that collocated sensor-actuator pairs offer a robust active feedback control system. Accordingly, consider a linear feedback system producing active forces at certain locations from the motion signals at the same locations:

$$f_l^{act} = C_l(\omega) \cdot u_l, \qquad (9.8)$$

where $C_l(\omega)$ are complex-valued control functions. Equations (9.7) and (9.8) form a system of linear equations governing the motion. Dynamic compliance at an arbitrary point $n$ can be represented by the familiar modal expansion:

$$G_{mn}(\omega) = \sum_k \frac{\varphi_{nk}\varphi_{mk}}{\omega_k^2(1 + i\eta_k(\omega)) - \omega^2}. \qquad (9.9)$$

Here $\varphi_{nk}$ are components of the normal modes that are, for sake of simplicity, assumed real for a low-damped primary system, $\omega_k$ are "undamped" natural frequencies and $\eta_k$ are associated loss factors. As mentioned above, main resonance peaks present the primary concern in reducing flexural vibration of optical tables. Suppose the "open-loop" dynamic response is dominated by the $k$th normal mode in the vicinity of its resonance frequency. Then the compliance can be approximately represented by a single member of the series (9.9) in the vicinity of $\omega_k$:

$$G_{mn}(\omega) \approx \frac{\varphi_{nk}\varphi_{mk}}{\omega_k^2(1 + i\eta(\omega)) - \omega^2}, \quad \omega \approx \omega_k. \tag{9.10}$$

The linear system (9.7), (9.8) with the approximate expressions (9.10) substituted for dynamic compliances can be solved *explicitly* in the following form:

$$\frac{u_n}{u_n^{\text{passive}}} = \frac{1}{1 + \frac{i}{\eta_k\omega_k^2} \cdot \sum_{l=1}^{L} \varphi_{lk}^2 \cdot C_l(\omega)}, \quad \omega \approx \omega_k, \tag{9.11}$$

where $u_n^{\text{passive}}$ represents the solution in absence of active forces. Equation (9.11) leads to two important conclusions. First, the optimal phase of the complex-valued control functions $C_l(\omega)$ is $(-\pi/2)$ in the frequency range of interest. This had to be expected because damping is known to be the most efficient way of reducing resonance vibrations. Second, the right-hand side of (9.11) does not depend on measurement location $(n)$ or excitation point $(m)$. This suggests that a small number of collocated sensor–actuator pairs can damp resonance vibration, created by any source, throughout the structure, provided the modal shapes have substantial components at the locations of dampers. Based on the analysis of the typical resonance modes of anisotropic plates known in the literature [14, 15], it can be conjectured that two active dampers placed at two corners of a rectangular vibration-isolated platform such as an optical table would effectively reduce dominant resonance modes. This conjecture was confirmed by experiments on optical tables and breadboards. One such experiment had been presented in Example 4.1 of Chapter 4 illustrating the laser Doppler vibration measurement methods.

Practical implementation of active damping is constrained by conditions of stability and limitations of control electronics. An active damping system based on inertial sensors and electromagnetic actuators is described in [12]. Figure 9.7 shows the general layout of the system (known as SmartTable®). Two sensor–actuator assemblies are integrated into the structure of the optical table at two corners.

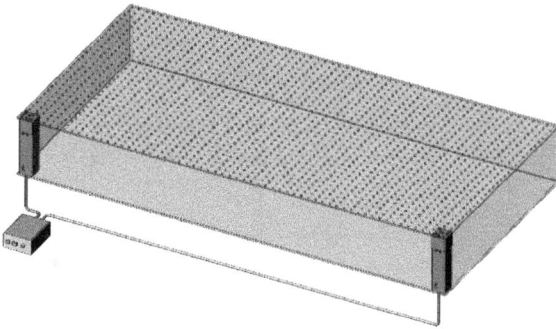

**Figure 9.7.** General layout of the SmartTable®. Adapted from [12].

The pre-amplifiers and band-pass filters condition the sensor output signal before digitizing and feeding it into the digital controller. The controller implements recursive filters that compensate for the dynamic response of the actuator and correct the phase to bring the phase difference between the displacement and force close to $(-\pi/2)$ in the desired frequency range. This, in effect, emulates the action of a viscous damper.

Figures 9.8–9.10 illustrate active vibration control performance under various vibration environments. Figure 9.8 compares the vibration spectra created by random excitation. The main resonance vibration peaks are reduced about tenfold. Figure 9.9 shows the dynamic compliances of the SmartTable obtained by applying random excitation (about 1.25 N RMS) through a force sensor. The damping performance is on par or better than that of top-of-the-line tables containing tuned vibration absorbers described in Section 7.3 of Chapter 7. However, if the table is loaded by a weight comparable to the weight of the table, the passive vibration absorbers can become "mistuned", whereas the active dampers work as well after auto-tuning (see [12] for the details of the auto-tuning process).

Figure 9.10 illustrates the effect of damping in the time domain. It shows the attenuation of vibration caused by a short triangular load pulse. The shock was applied near a corner about 25 cm from the edges of the table; the resulting vibration was monitored by an accelerometer near the center of the table. Attenuation to the background level happens about ten times faster when the active damping is switched on.

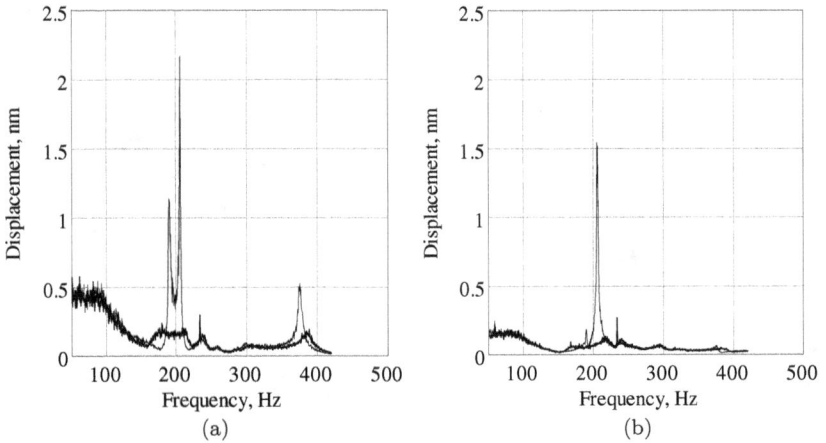

**Figure 9.8.** Linear spectra of vibration near the corner (a) and at the center (b) of the table. Thick line: active damping on; thin line: active damping off. Table size: 2.44 m × 1.22 m × 203 mm. Adapted from [12].

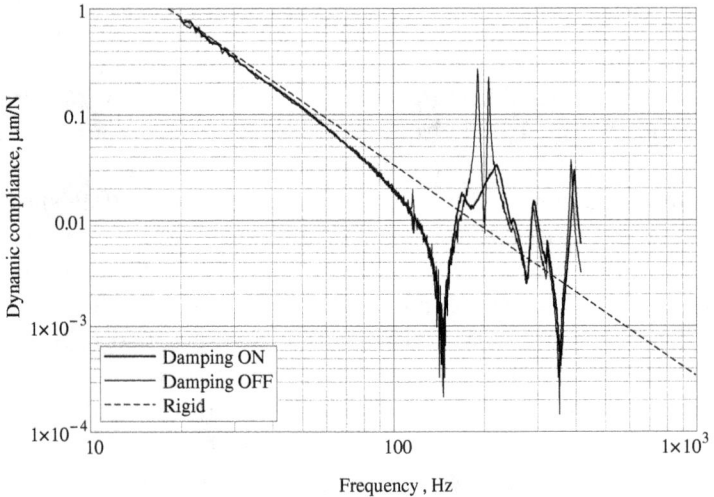

**Figure 9.9.** Dynamic compliance of an optical table with active vibration damping near a corner. Adapted from [12].

Further developments in the field of active damping, including add-on damping systems and spatial configurations, are reviewed in [16].

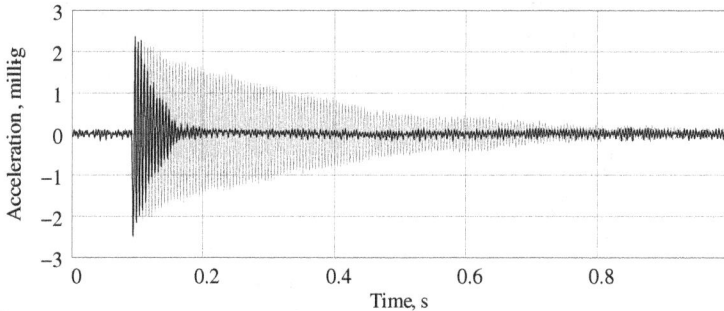

**Figure 9.10.** Vibration attenuation of an optical table with active vibration damping. Dim line — active control off, dark line — active control on. Adapted from [12].

## 9.3 Unique Vibration Control Systems

The previous two sections describe examples of active vibration control solutions that were developed as standard products and can be manufactured on an industrial scale for a wide circle of users. However, as science probes the frontiers of achievable, unique requirements arise that demand unique vibration control solutions. These solutions are less vulnerable to cost constraints and other limitations. This section is a brief review of two such systems: vertical vibration isolation for precision atom interferometry measurements by Hensley *et al.* [17], and 12-axis seismic active isolation for LIGO by Matichard and co-authors [18,19]. These two systems have been chosen for review because they achieve isolation performance far exceeding that of commonly used commercial systems.

### 9.3.1 *Vertical vibration isolation for the atom interferometer*

Atom interferometry is a relatively new area of experimental physics. It is based on the observation that matter, like light, can exhibit wave properties. This technique is used for precise measurements of gravitational acceleration [17, 20]. The authors note that some applications in geophysics, metrology, and geology require measuring accuracy ranging from $10^{-6}$ g to $10^{-9}$ g. They state, furthermore, that to achieve this level of precision with their measurement method "we need to stabilize the position of a mirror to a small fraction of the wavelength of light for times approaching 1 s". Without isolation, typical floor vibration in the 0.1–10 Hz range would "completely wash out the interferometric fringes". The authors estimated that vibration isolation system for this experiment should have a resonance frequency substantially below 0.1 Hz. The interferometer is mainly sensitive to vertical vibration; therefore, it was surmised that a single-axis (vertical) low-frequency isolation would be sufficient for a major reduction of the measurement noise. The isolation system is described in [17]; the paper by Peters *et al.* [20] presents a slightly different design and contains a detailed description of the application.

The principal schematic of the isolation system is shown in Fig. 9.11. The optical element to be protected (the mirror), together with the vibration sensor (accelerometer) is suspended on soft helical springs; damping material is attached to the sides of extended springs to suppress the internal resonances. An air piston is used to ensure uniaxial frictionless motion in the vertical direction: the inner cylinder is attached to the mirror assembly, and the outer cylinder, divided from the inner cylinder by a small air gap, is attached to a relatively stiff support structure. The whole assembly is placed on an optical table floating on pneumatic vibration isolators.

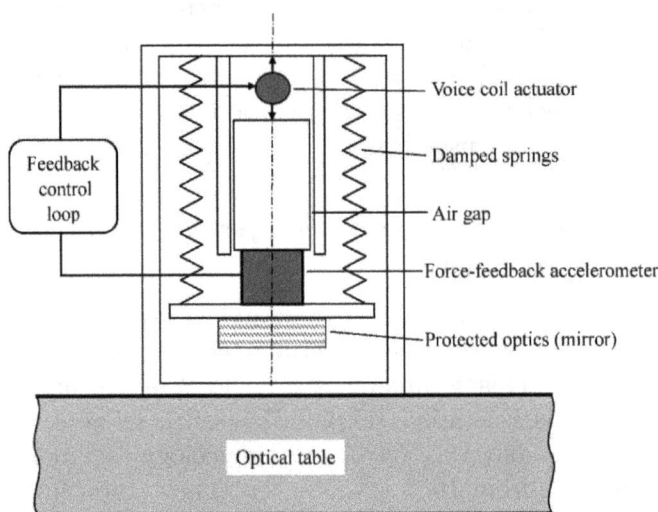

**Figure 9.11.**  Schematics of the unidirectional active vibration isolation system.

The optical table and the springs form a two-stage passive isolation system. The partial frequencies of the floated table and the spring suspension are both approximately 1.6 Hz, which makes the combination close to an optimal one as described in Section 3.4 of Chapter 3. Passive isolation alone reduces vibrations at frequencies above 3 Hz, but at the expense of some amplification at 1.2 Hz and 1.8 Hz (resonance frequencies of the resulting 2DOF system). To achieve the required level of vibration isolation in the 0.1–10 Hz frequency range, an active system was necessary.

A special type of vibration sensor, a force-feedback accelerometer, was employed to obtain extremely low-noise low-frequency vibration signal. It is an inertial sensor with seismic mass suspended from a base on a soft spring, much as that of a geophone (Section 4.1 of Chapter 4). The motion of the mass is monitored by a capacitive position sensor; however, instead of using this signal directly, it is fed into a feedback loop that drives a linear force transducer; this feedback is designed to cancel the motion of the mass. The feedback voltage measures the force necessary to stop the mass, and therefore the base acceleration. The authors report that the particular model of the sensor they used achieved sensitivity of approximately 2000 V/(m/s$^2$) and noise level below $4 \cdot 10^{-9}$ (m/s$^2$)/Hz$^{1/2}$ (much lower than piezoelectric accelerometers considered in Section 4.1 of Chapter 4) in the 0.1–10 Hz frequency range.

Exceptional care was taken to align the axis of the air piston, the sensitive axis of the accelerometer, the line of action of the actuator, the central axis of the protected optics, and the normal to the surface of the mirror along the same vertical axis.

The control algorithm comprises negative acceleration and velocity feedback loops with additional filters required for stability (compensating for the finite bandwidth of the sensor). The feedback signal drives a voice coil actuator placed between the support structure and the isolated optics–sensor–air cylinder assembly. Additional improvement was contributed by active control of the rocking modes of the optical table. The test results demonstrate vibration isolation starting from 0.01 Hz and reaching a factor of 300 in the critical frequency range, thereby reducing the vibration of the protected optics to the level of approximately $10^{-8}$ (m/s$^2$)/(Hz$^{1/2}$).

This design uses electromagnetic actuators and soft springs, and therefore belongs to the class of soft (parallel) active isolation systems as described in Section 9.1. Exceptional performance was achieved by using a super low-noise low-frequency vibration sensor and taking advantage of the largely uni-directional nature of the vibration problem.

### 9.3.2 *Seismic isolation for detection of gravitational waves*

On September 14, 2015, for the first time ever, Laser Interferometer Gravitational-Wave Observatory (LIGO) interferometers detected gravitational waves, thereby confirming the major prediction of Einstein's general relativity theory made a century ago, and opening new horizons in the research into the nature of our Universe. This was a triumph of astrophysics, optics, and optoelectronics. It was also a triumph of the technical discipline of vibration control. Professor Weiss, one of the leading authors of the gravitational waves experiment, speaking at the NSF press conference [21], singled out the problem of vibration isolation as the most important technical hurdle on the way to the successful observation of gravitational waves by LIGO. A general description of the experiment can be found in the paper [22]. Here we'll give a brief review of the vibration control system following the papers [18, 19]. According to the authors, the resulting vibration isolation system, called Advanced LIGO seismic isolation system, was a result of over a decade of research and development by a large group of contributors.

The advanced LIGO detector is a modified Michelson interferometer that measures gravitational-wave strain as a difference in the lengths of its orthogonal arms. Its principle of action is further explained in [22]: "Each arm is formed by two mirrors, acting as test masses, separated by $L_x = L_y = L = 4$ km. A passing gravitational wave effectively alters the arm lengths such that the measured difference is $\Delta L(t) = \delta L_x - \delta L_y = h(t)L$, where $h$ is the gravitational-wave strain amplitude projected onto the detector. This differential length variation alters the phase difference between the two light fields returning to the beam splitter, transmitting an optical signal proportional to the gravitational-wave strain to the output photodetector."

The required high sensitivity of the system dictated unprecedented degree of isolation from the ground vibration (seismic noise).

Each test mass is suspended on a quadruple (four-stage) pendulum system supported by an active vibration isolation platform. All components except the laser source are mounted on vibration isolation stages in ultrahigh vacuum. Contrary to the problem solved in [17, 20], vibration isolation of LIGO faced the problem of controlling all 6-DOF of vibration of the core optics (the test masses and the beamsplitter) and the auxiliary optics.

Advanced LIGO contains three active isolation sub-systems. The first subsystem consists of hydraulic platforms called Hydraulic Exovacuum Pre-Isolators (HEPI) that are used as pre-isolators outside of the vacuum chambers. The second subsystem includes five single-stage active platforms (Intra-vacuum Seismic Isolators) that are mounted on the HEPI and support the interferometer's auxiliary optics. They are located inside the vacuum chambers called Horizontal Access Modules (HAM-ISI). Finally, the interferometer uses five Basic Symmetric Chambers–Intra-vacuum Seismic Isolators (BSI-ISI) units to support the core optics: one for each of X-arm test masses, one for each of Y-arm test masses, and one for the beamsplitter. The design requirement for isolation performance of BSI-ISI was to provide more than three orders magnitude of isolation, thereby bringing the motion of core optics down to $10^{-11}$ m/Hz$^{1/2}$ at 1 Hz and $10^{-12}$ m/Hz$^{1/2}$ at 10 Hz.

ISI active platforms use soft springs and electromagnetic (voice coil) actuators, and therefore belong to the class of soft (parallel) active isolation systems as described in Section 9.1; they achieve exceptional performance due to the use of relative displacement sensors and feedforward sensors in addition to inertial sensors in the feedback loop, and intricated control algorithm that allows for reduction of sensor noise.

A schematic diagram of core optics supported by HEPI and BSI-ISI according to [19] is shown in Fig. 9.12. The test mass is suspended on the quadruple pendulum that is supported by two optical tables in a two-stage configuration; each table is equipped with a complete 6-DOF active vibration isolation system. This structure, in turn, rests on the hydraulic active isolation platform. HAM-ISI isolation system has the same basic architecture but contains only a single-stage optical table.

Figure 9.13 shows a schematic representation of the sensor–actuator arrangement of one stage of the active isolation system.

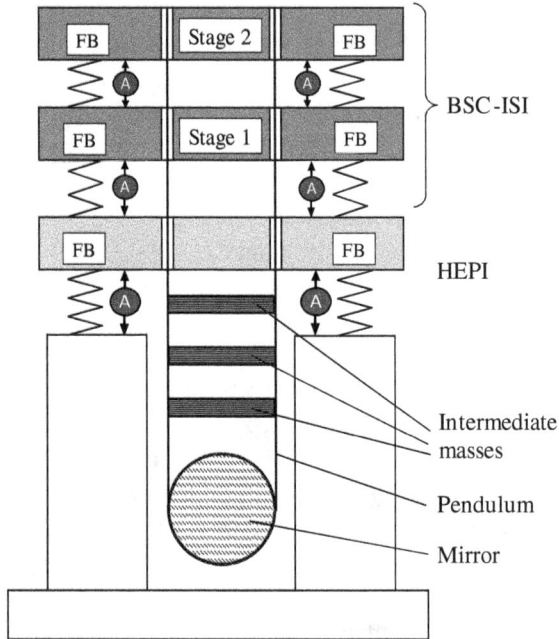

**Figure 9.12.** Core optics is supported by seven stages of isolation, three of them active. FB — feedback inertial sensors, A — voice coil actuators. Six sensor–actuators pairs are used on Stage 1 and Stage 2, as well as on the HEPI platform, to cover all principal rigid-body degrees of freedom. Additional sensors (relative motion and feedforward) are not shown. See more details in Fig. 9.13.

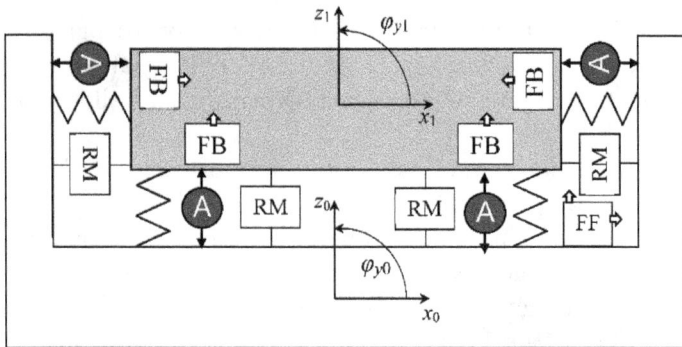

**Figure 9.13.** One stage of the active isolation system includes six inertial feedback sensors (FB), six electromagnetic actuators (A), six relative motion sensors (RM), and three 3-DOF inertial feedforward sensors (FF).

Six inertial sensors (three vertical and three horizontal) are used to monitor the absolute motion of the stage, and six relative motion sensors are used to measure the relative displacement. Similarly, six actuators are employed to apply the control forces. The sensor signals are combined via matrix transformation to measure the motion in Cartesian coordinates in the basis aligned with the axis of the interferometer (rigid body translations and rotations, see Fig. 9.13). Accordingly, the actuators are driven to provide the required forces and torques in the Cartesian basis. Decentralized control algorithm was used, with each control channel dedicated to a particular rigid-body degree of freedom. Geophones (see Section 4.1 of Chapter 4) were used as inertial vibration sensors.

The control algorithm is explained as follows [19]. Consider 1-DOF, say, $X$. Denote the input (base) signal $x_0$ and the output (table) signal $x_1$. The control force is $F$. Then

$$x_1 = P_S x_0 + P_F F.$$

Here $P_S$ is seismic (passive) transmissibility, $P_F$ is the transfer function of the control force. Absolute motion is measured with geophones that roll-off, and therefore are noise-limited, at low frequencies (below 100 mHz). To correct for this shortcoming, relative motion sensors are used at low frequencies to measure the differential motion $(x_1 - x_0)$. To reduce the impact of low-frequency noise, the absolute motion signal $x_1$ is filtered through a high-pass filter $H$, and the relative motion signal $(x_1 - x_0)$ is filtered through a low-pass filter $L$. These two filters are complementary ($H + L = 1$ at all frequencies) and tuned to obtain a suitable trade-off between active isolation at "high" frequencies (above 100 mHz) and noise amplification at "low" frequencies (below 100 mHz). The blended signal is fed to the feedback controller $C_{\mathrm{FB}}$:

$$F = -C_{\mathrm{FB}}[H x_1 + L(x_1 - x_0)].$$

The resulting transmissibility is given by

$$\frac{x_1}{x_0} = \frac{P_S + L C_{\mathrm{FB}} P_F}{1 + C_{\mathrm{FB}} P_F}. \tag{9.12}$$

Additional improvement of transmissibility is based on using the signal from the inertial sensor monitoring the ground motion. It is filtered by the sensor correction filter $H_{SC}$ and added to the signal from the relative motion sensor. The goal is to limit the noise from the relative motion sensor. A single instrument is used for sensor correction of multiple platforms, which is important for interferometers sensing differential motion. As a result, the equation for $F$ takes the form

$$F = -C_{\text{FB}}[Hx_1 + L(x_1 - x_0) + LH_{\text{SC}}x_0],$$

and the transmissibility is expressed as

$$\frac{x_1}{x_0} = \frac{P_S + LC_{\text{FB}}P_F(1 - H_{\text{SC}})}{1 + C_{\text{FB}}P_F}. \tag{9.13}$$

Equations (9.12) and (9.13) are sophisticated analogs of Eq. (9.6) in this advanced control scheme. Details of the filters can be found in [19].

The suspension springs are composed of flexible rods for horizontal compliance and leaf springs (blades) for vertical compliance. The blades have a triangular shape so that, under the end load, they assume a constant radius of curvature [18] and therefore almost uniform stress according to Bernoulli–Euler beam theory (see Section 3.5 of Chapter 3).

Although the main attention in the design of the vibration control system was directed to rigid-body modes of stages, flexural vibrations of various structural sub-systems needed to be addressed. It is noted in [18] (Part 2) that the main structural resonance of Stage 1 of BSI-ISI was identified at 220 Hz, and dynamic vibration absorbers (see Section 7.2 of Chapter 7) were used to suppress it. Another set of dynamic absorbers was installed on the payload frame supporting the optics to suppress the resonance near 100 Hz. The authors note that this passive damping simplified the implementation of active control and improved its robustness.

The BSI-ISI system provides isolation starting from 0.1 Hz and reduces vibration by three orders of magnitude in the entire range from 1 to 30 Hz. Additional isolation is provided by the one-stage HEPI based on hydraulic actuators. As a result, the system meets the design requirement of bringing the motion of core optics down to $10^{-11}$ m/Hz$^{1/2}$ at 1 Hz and $10^{-12}$ m/Hz$^{1/2}$ at 10 Hz. Design and implementation of the Advanced LIGO seismic isolation system combines and advances best practices of vibration control to achieve spectacular performance.

# References

1. C.R. Fuller, S.J. Elliott, and P.A. Nelson, *Active Control of Vibration*, AP, London, 1996, 332 p.
2. A. Preumont, *Vibration Control of Active Structures: An Introduction*, Springer, Netherlands, 2014, 368 p.
3. M.D. Genkin, V.G. Elezov, and V.V. Yablonsky, *Methods of Controllable Vibro-Protection for Machinery* (in Russian). Nauka, Moscow, 1985, 240 p.
4. M.D. Genkin and V.M. Ryaboy, *Elastic-inertial Vibration Isolation Systems: Limiting Performance, Optimal Configurations* (in Russian), Nauka, Moscow, 1988, pp. 52–78.
5. V.M. Ryaboy, Limiting performance and optimal synthesis of the vibration isolating structures, *Structural Dynamics: Recent Advances. Proceedings of the 5th International Conference*, Southampton, vol. 2, pp. 681–688, 1994.
6. V.M. Ryaboy, Limiting performance estimates for active vibration isolation in multi-degree-of-freedom mechanical systems, *Journal of Sound and Vibration*, vol. 186, no. 1, pp. 1–21, 1995.
7. V.M. Ryaboy, Limiting performance estimates for active vibration isolation in free-free multi-degree-of-freedom mechanical systems, *Proceedings of the $68^{th}$ Shock and Vibration Symposium*, SAVIAC, 1997, pp. 685–691.
8. A. Beard, D. Schubert, and A. Von Flotow, A practical implementation of an active/passive vibration isolation system, *Proceedings of the SPIE Vol. 2264: Vibration Monitoring and Control*, 1994, pp. 38–49.
9. E.H. Anderson and B. Houghton, ELITE-3 active vibration isolation workstation, *Proceedings of the SPIE Vol. 4332, SPIE Smart Structures and Materials 2001: Industrial and Commercial Applications of Smart Structures Technologies*, 2001, pp. 4332–4323.
10. *Guardian*$^{TM}$ *Workstation. User's Manual.* Newport Corporation, 2016, p. 23. https://www.newport.com/medias/sys_master/images/images/ha1/hc3/8922102169630/AI-6-Guardian-Worksation-Users-Manual-revA.pdf. Accessed on March 1, 2021.
11. L.P. Fowler, S. Buchner, and V. Ryaboy, Self-contained active damping system for pneumatic isolation tables, *Proceedings of SPIE Vol. 3991, Industrial and Commercial Applications of Smart Structures Technologies*, ed. by J. Jacobs, 2000, pp. 261–282.
12. V.M. Ryaboy, P.S. Kasturi, A.S. Nastase, and T.K. Rigney, Optical table with embedded active vibration dampers (Smart table), *Proceedings of SPIE Vol. 5762, Smart Structures and Materials 2005:*

*Industrial and Commercial Applications of Smart Structures Technologies*, ed. by E.V. White, pp. 236–244.

13. Y.-S. Lee, P. Gardonio, and S.J. Elliott, Coupling analysis of a matched piezoelectric sensor and actuator pair for vibration control of a smart beam, *Journal of the Acoustical Society of America*, vol. 111, no. 6, pp. 2715–2726, 2002.

14. M.R. Maheri and R.D. Adams, Modal vibration damping of anisotropic FRP laminates using the Rayleigh-Ritz energy minimization scheme, *Journal of Sound and Vibration*, vol. 259, no. 1, pp. 17–29, 2003.

15. R.D. Adams and M.R. Maheri, Natural frequencies and damping in fibre reinforced plastic plates, *Proceedings of the 5th International Conference on Recent Advances in Structural Dynamics*, vol. 2, University of Southampton, UK, 1994, pp. 699–713.

16. V.M. Ryaboy and P.S. Kasturi, Active sensor/actuator assemblies for vibration damping, compensation, measurement and testing, *Proceedings of the SPIE Vol. 7643, Active and Passive Smart Structures and Integrated Systems*, 2010, ed. by Mehrdad N. Ghasemi-Nejhad, 76431J, doi:10.1117/12.847331.

17. J.M. Hensley, A. Peters, and S. Chu, Active low frequency vertical vibration isolation, *Review of Scientific Instruments*, vol. 70, no. 6, pp. 2735–2741, 1999.

18. F. Matichard *et al.*, Advanced LIGO two-stage twelve-axis vibration isolation and positioning platform. Part 1: Design and production overview, *Precision Engineering*, vol. 40, pp. 273–286, 2015; Part 2: Experimental investigation and test results, *ibid.* pp. 287–297.

19. F. Matichard *et al.*, Seismic isolation of Advanced LIGO: Review of strategy, instrumentation and performance, *Classical and Quantum Gravity*, vol. 32, no. 18, 185003, 2015.

20. A. Peters, K.Y. Chung, and S. Chu, High-precision gravity measurements using atom interferometry. *Metrologia*, vol. 38, no. 1, pp. 25–61, 2001.

21. https://www.ligo.caltech.edu/video/ligo20160211v11; https://www.ligo.caltech.edu/WA/video/ligo20160211v12. Accessed on March 1, 2021.

22. B.P. Abbott *et al.*, Observation of gravitational waves from a binary black hole merger, *Phys. Rev. Letters*, PRL 116, 061102, pp. 061102-1–061102-16, 2016.

# Conclusion

The requirements to vibration control of optomechanical systems are constantly increasing. This trend will continue in the future. Although the principles of vibration control have been known for decades, innovative designs and products are appearing and becoming available all the time. Solid knowledge of fundamental principles and methods, which this book attempted to provide, is necessary for confident orientation in the ever-changing field of vibration control. This knowledge is a guide to finding adequate solutions to challenges of demanding applications, be it on an early design stage or during a later design or system integration steps, addressing unexpected vibration problems.

The earlier in the design stage vibration control is taken into consideration, the better. Reducing vibrations at their source is the best way of vibration control, but it is not always feasible. Vibration isolation and vibration damping, implemented in various vibration control devices or in the optomechanical application itself, are two techniques for reducing undesirable vibration effects.

Vibration-resistant design has its limitations and trade-offs. For example, internal means of damping (integral damping) reduce resonance reaction of the structure, but their principle of work requires relative motion; add-on devices such as constrained damping layers and dynamic absorbers add mass and require precision frequency tuning.

Vibration isolation in form of optical tables, workstations, or individual isolators is ubiquitous in state-of-the-art optomechanical systems and will serve immediate needs of precise experiments and manufacturing processes in the future, although it has its inherent limitations.

Active vibration control, which is gaining increased acceptance, allows for unprecedented levels of vibration immunity, thereby enabling new scientific and technological breakthroughs.

# Index

www.ingramcontent.com/pod-product-compliance
Lightning Source LLC
Chambersburg PA
CBHW050548190326
41458CB00007B/1969

www.ingramcontent.com/pod-product-compliance
Lightning Source LLC
Chambersburg PA
CBHW050548190326
41458CB00007B/1969